Mammals

300 Amazing Animals

Mammals
300 Amazing Animals

Chris McNab

Reprinted in 2016, 2021

Revised edition published in 2016

Copyright © 2006 Amber Books Ltd.

First published in 2006

All rights reserved. No part of this publication may be reproduced, stored in a retrieval system, or transmitted in any form or by any means, electronic, mechanical, photocopying, recording, or otherwise, without prior written permission of the copyright holder.

Published by
Amber Books Ltd
United House
North Road
London N7 9DP
United Kingdom
www.amberbooks.co.uk
Instagram: amberbooksltd
Facebook: amberbooks
Twitter: @amberbooks

ISBN: 978-1-78274-385-9

Project Editor: Sarah Uttridge
Design: Graham Curd

Printed in China

Picture Credits:
TRH Pictures

Artwork Credits:
All artworks © International Masters Publishing BV except the following:

7: National Oceanic and Atmospheric Administration/Department of Commerce; 8: U.S. Fish and Wildlife Service; 10: Image Library Gold; 11: U.S. Fish and Wildlife Service; 13: Backgrounds Archive.

CONTENTS

Introduction	6
Artiodactyla–Carnivora	14
Cetacea–Insectivora	149
Lagomorpha–Pinnipedia	182
Primates–Rodentia	230
Sirenia–Xenartha	304
Glossary	312
Index	314

Introduction

In the arid grasslands of sub-Saharan Africa lives *Petromyscus collinus*, the pygmy rock mouse. Under equatorial temperatures of 40°C (105°F), and surrounded by a world of predators, this tiny creature manages to scrape out an existence, despite weighing about 25g (less than an ounce) and measuring, including tail, an average of 16cm (6in). Its diet is mainly seeds and occasional insects. A female pygmy rock mouse can produce a litter of two to five young after a gestation period of only 30–33 days, the naked and helpless infants feeding on their mother's teats for a few weeks before being dispersed into the wilderness to survive on their own.

Now let's leave the pygmy mouse's habitat, and travel to its opposite extreme – the ice-choked Arctic seas off eastern Greenland. Here is found one of the world's largest animals, *Physeter catodon*, the sperm whale (which is also found in many other global waters, tropical as well as Arctic). The sperm whale is an aquatic monster, a creature growing to 20m (65ft) in length and weighing 57 tons (63 tonnes). Living on squid, octopus, fish and other marine life, it can dive to a depth of 3000m (9842ft) – most submarines would be crushed at depths well within 1000m/3280ft. After a 14- to 15-month gestation, a female gives birth to a single infant that is 4m (13ft) long and initially dependent on the mother, although fully formed and active.

To make a direct association between the two animals described above may seem a verbal contrivance, however, both animals belong to the class *Mammalia*, a category that unites 4475 species of prodigious diversity, from polar bears to flying squirrels, bats to anteaters, mink to dolphins. Binding them all is a set of common physical traits that distinguishes them from the rest of the animal world. It is these traits that have not only allowed mammals to inhabit every corner of the planet, but have also given one of their species – humans – a position of total dominance over all other creatures.

MAMMAL CHARACTERISTICS

All mammals are vertebrates, and despite physiological differences they have some anatomical constants. Mammals have, to greater or lesser degrees, external hair. This may not be readily apparent. The naked mole rat, for example, appears to have totally bare skin, although it actually has a thin layer of extremely fine hairs. By contrast, the sea otter has 150,000 hairs per

The blue whale is the largest living animal in the world. The main arteries of this mammal are so large that a small person could crawl through them.

square centimetre (1 million per square inch). Hair coverings control heat loss and provide protection for the skin, but they also have a role in social interaction (such as species identification and mating displays), sensory detection (in the form of whiskers) and even defence – porcupine spines are, in fact, rigid hairs. Beneath the hair, mammals have a common skin structure, consisting of an outer protective skin layer known as the 'epidermis', and an inner layer, the 'dermis', which contains blood vessels, glands and nerve endings. The skin has an active role in temperature regulation and it also produces the scents and sweats by which many mammals identify one another and mark out territories.

Internally, mammals share other connections. Mammalian ears contain a unique three-bone structure – the malleus, incus and stapes – which works to amplify sound waves and direct them to the eardrum. Note also that mammals have external ears located on the outside of the head. These serve to funnel sounds into the inner ear, and many mammals have directional control over them. Mammals have a single jawbone hinged directly to the skull, something that gives them great power in the bite – an African hyena can break through large bones. Furthermore, mammals have a unique group of teeth structures with individual purposes: incisors for biting and stripping, canines for killing and gripping, and molars and premolars for grinding up food for swallowing. The combination of these teeth structures varies according to the mammal. Indeed some species have entirely different oral configurations – many species of whale have specialized baleen plates for sieving food from the water.

REGULATING BODY TEMPERATURE

A fundamental connection between mammals is that they are endothermes – they maintain their core body temperature within a relatively narrow and constant range, the heat produced being a by-product of food digestion. In

most mammals except marsupials, the temperature range is between 36°C (96.8°F) and 40°C (104°F); marsupials have a range of 27°C (80.6°F) to 32°C (89.6°F). Maintaining the core temperature is vital, as movement either above or below can be life-threatening. Core temperature is controlled through the hypothalamus in the brain, but there are various physiological and behavioural activities designed to conserve heat or regulate heat loss. To conserve heat, the mammal can shiver (to generate muscular heat), share body heat with other animals, find places of shelter, puff up fur to trap more warm air, or eat, while the blood vessels on the skin will contract to reduce the amount of warmth being lost through the skin surface and extremities. By contrast, heat loss is accomplished by dilating the surface blood vessels, panting (to expel hot air), opening the body posture to maximize the heat loss area, or submerging or standing in water.

Some mammals hibernate during winter, their bodies almost going into a state of suspension. Hibernation takes place in a warm den or nest, and the animal's body temperature will plunge – a European hedgehog's hibernation temperature is about 6°C (43°F). Breathing cycles and metabolic rate are also reduced, but the animals will revive with no harmful effects in the summer.

REPRODUCTION

A defining quality of mammals is that they give birth to live young (with the exception of egg-laying echidnas and the platypus), and the young are fed on milk from the female's mammary glands. Reproduction begins with an act of internal fertilization. After this, the infants develop inside

Sea otters spend up to 48 per cent of the daylight hours grooming their dense fur. They groom themselves by rubbing fur with their forepaws.

African elephants consume about six to eight per cent of their body weight in vegetation each day, sometimes feeding for 18 hours per day.

the mother over a period ranging from a bottom limit of 12 days for some mouselike mammals up to 22 months for the African elephant. Newborn mammals vary in their levels of activity and dependence. A kangaroo's joey, for example, spends 190 days in its mother's pouch before venturing into the outside world, whereas a wildebeest can walk within 15 minutes of birth. Small mammals such as rodents produce the largest litter sizes, up to 20 offspring in some species, while the large placental mammals (meaning that the developing foetus is fed nutrients and oxygen through a placenta) tend to have only one or two young, which are more highly developed.

ADAPT AND SURVIVE

Endothermic bodies have meant that mammals have adapted to all of the earth's environments, an achievement shared only with birds. Consequently, each species type has had to develop unique skills of survival, adapting itself to the habitat both physically and behaviourally.

The practicalities of mammal existence are fairly few – the creature must eat and drink, it has to survive climatic adversity, it must avoid being killed by predators, and, for the sustainability of the species, it must reproduce. Mammals are typically classified as carnivores (they eat other animals only), herbivores (they eat plant matter only) or omnivores (they eat both). Generally speaking, the largest land mammals are herbivores, as they lose heat slowly on account of their size and therefore can afford the extended digestive process of a plant diet. Mid-sized and small mammals have more rapid heat loss, and therefore eat meat (as part of a carnivorous/omnivorous diet), which is broken down more rapidly. Fluids are obtained by drinking or are extracted from the blood of prey or, in the case of some animals, are extracted almost entirely from plant matter.

Many land mammals, particularly smaller creatures such as rodents and marsupials, enhance their chances of survival by creating nests, burrows or dens. These serve a number of important functions. First, they provide shelter

There may be only about 500 monk seals left in the world. This makes them the most endangered species of seal.

from the climate and a place to sleep in (relative) safety from predators. They are also social areas, often being the home of a family/social unit or the psychological centre of a territory. Some mammal shelters can be very elaborate constructions indeed. The beaver's lodge, for example, is a complex nest of mud-bound sticks accessed by underwater entrances, and the beaver can control the water flow and water level around the lodge by creating dams up to 32m (100ft) long.

SOCIAL LIFE

In the human world, social interaction is mainly about establishing status in relation to those around us. The same applies to other mammals, although the success of their interaction has a fundamental relationship to physical, instead of emotional, survival. Mammals have an enormous range of social structures. Many mammals are solitary, interacting with others only during mating or when a female has offspring. The others are group animals, but group structure varies hugely. Some groups may be no more than a male – female pair in a monogamous relationship shared only with their immediate offspring. By contrast, Mexican free-tailed bats live in colonies that may number up to 20 million animals. Groups are often hierarchical in nature, particularly among males, with certain animals achieving sexual dominance or feeding dominance through their physical prowess. Groups can also separate along gender lines, interacting only when breeding – seals and elephants are typical examples. Its place in the social order has a major effect on a mammal's reproductive potential, particularly among males, who will often literally do battle to win access to a fertile female.

A common part of mammal social behaviour is territorialism. Territories can be as little as a few square metres to several hundred square kilometres. These are marked out by several means – scent-gland emissions, urine, faeces, piles of mud and even noise (often in the case of primates). The mammal may defend a territory violently when a rival animal enters, particularly if that animal is there in an attempt to steal food, acquire breeding rights or assume social dominance. Although territorialism may seem like an example of simple animal aggression, for many species it is vital to protecting food and family, and therefore is intrinsic to survival.

This book contains 299 of the world's mammals, a fraction of the total species. (They are arranged here by order, then family, then alphabetically by Latin species name within the family.) Unfortunately, a large percentage of all mammals are now threatened with extinction due to human activity –

Giant pandas are among the most recognized but rarest animals in the world. They have come to symbolize endangered species.

hunting (for fur, food or just pleasure), deforestation and pollution. Many species have become extinct over the past 200 years, and we may destroy many more. Primates, whales and big cats are particularly threatened, but most other mammal families have endangered species. As an example of human activity, the population of North American buffalo was hunted from 60 million animals in the late eighteenth century to fewer than 1000 animals by the end of the nineteenth century. In another 100 years, mammals such as tigers, polar bears, pandas and blue whales may exist only on the pages of books or in zoos. We must realize, and quickly, that stripping the world of its wildlife and habitat may ultimately have a catastrophic effect on us, the world's most successful mammal, now possibly heading for self-destruction.

ARTIODACTYLA: ANTILOCAPRIDAE

Pronghorn

The pronghorn is the only existing member of the *Antilocapridae* family and is physically unique. Its horns are structured like an antelope's, featuring a sheath of horn over a core of bone, but are forked and shed annually in the manner of deer antlers. Female horns are shorter than the ears, but male horns can grow up to 50cm (20in). The animal is also one of the fastest creatures on the planet. It can run at, and maintain, speeds of up to 65km/h (45mph). Even a two-day-old fawn can achieve a pace of around 40km/h (25mph). Keen eyesight enhances the pronghorn's defence against predators, and it is able to detect movement at a distance of 4km (2.5 miles) over open terrain. Pronghorns have a red-brown or tan coat with a white underside, the male featuring a black neck patch. They are durable creatures, able to survive in both arctic and tropical conditions, moving about in small groups during the summer and large herds during the winter. They feed on grasses, shrubs and even cacti.

Species name:	*Antilocapra americana*
Features:	Red-brown to tan coat with white underside; male has black neck patch and horns longer than ears
Habitat:	Grasslands, plains and brushlands
Distribution:	West and central North America
Length:	Up to 1.5m (4ft 11in) without tail
Weight:	Up to 70kg (155lb)
Breeding:	Typically twin infants born after a 250-day gestation

ARTIODACTYLA: BOVIDAE

Addax

The addax is one of the world's most imperilled mammals, and has been given a 'critically endangered' status by the International Union for the Conservation of Nature (IUCN). Hunting and the expansion of industrial and agricultural interests into its habitats have been the primary reasons for the addax's decline. Because of its rarity, addax herds seldom consist of more than four animals, and many addax lead entirely solitary lives. The addax is, however, superbly adapted to the desert conditions of Saharan North Africa. Its hooves are wide and well adapted for walking on soft sand, and it rarely needs to drink – most of its fluids are extracted from hardy desert grasses, herbs and acacia plants. The addax conserves its body fluids by searching for food during the twilight periods or at night, resting up in shade during the daytime. If a pair finds the opportunity to mate, gestation lasts for around 300 days. A female usually gives birth in the period from September to January, after the rainy season.

Species name:	*Addax nasomaculatus*
Features:	Grey-brown coat in winter, moving to sandy or white in summer; chestnut forehead tuft; spiral horns
Habitat:	Arid desert regions, particularly dune areas
Distribution:	Northwest Africa
Length:	Up to 1.7m (5ft 6in) without tail
Weight:	Up to 125kg (280lb)
Breeding:	One calf born after a 300-day gestation

ARTIODACTYLA: BOVIDAE

Impala

The impala is a hardy native of East and southern Africa. It is a medium-sized antelope species, generally growing to 71–91cm (28–36in) in height at the shoulder, and is particularly known for the leaping zigzag run it makes when escaping from predators. When conducted by a massed herd, the running pattern makes it extremely difficult for a predator to target a specific individual. Visually, the impala has reddish-fawn fur with black lines on the tail, lower back and thighs. They require daily access to drinking water, so are generally found near rivers and waterholes. Impalas form themselves into mixed male/female herds during the dry season, when scarcity of food and water precludes the territorial behaviour of males common during times of plenty. When food supplies are stable, small groups of six to eight male impalas will mark out territories of about 7.7 sq. km (3 sq. miles), using urine, dung and emissions from face glands. Lengthy horn duels are conducted between males vying for dominance, the powerful horns occasionally inflicting serious injury. The impala diet consists of grass shoots and leaves.

Species name:	*Aepyceros melampus*
Features:	Reddish-brown coat accented by black lines; scent gland above heel on hind legs
Habitat:	Savannah and woodlands
Distribution:	East and southern Africa
Length:	Up to 1.5m (5ft) without tail
Weight:	Up to 65kg (145lb)
Breeding:	One fawn after a 194- to 200-day gestation

ARTIODACTYLA: BOVIDAE

Springbok

The springbok was the national symbol of South Africa for much of the country's recent history. Springboks are small – they rarely exceed about 1.4m (4ft 6in) in length and are around 76cm (30in) tall at the shoulder – and they have a reddish-brown upper coat and white underparts. When alarmed or excited, springboks put on impressive displays of 'pronking' (from the Afrikaans meaning 'to show off') – making a leaping run, they bounce off the ground with rigid legs and an arched back to heights of up to 4m (12ft). This form of run is not designed for speed; when a springbok is directly chased by a predator, it will adopt a more conventional run at speeds of up to 64km/h (40mph). Instead, the pronk seems designed to display physical fitness, thereby deterring predators from giving chase in the first place, and possibly confusing the lines of attack. Suitably for the climate, springboks have excellent water conservation capabilities, and during acute dry spells can live purely off the water extracted from foliage.

Species name:	*Antidorcas marsupialis*
Features:	White underparts; red-brown band on the face; both sexes have horns
Habitat:	Grasslands and arid regions
Distribution:	Southern Africa
Length:	Up to 1.4m (4ft 6in) without tail
Weight:	Up to 48kg (105lb)
Breeding:	One calf after a 194- to 200-day gestation

ARTIODACTYLA: BOVIDAE

Blackbuck

The blackbuck is a powerful gazelle native to the Indian subcontinent. Extremely long twisted horns are the key feature of adult males, and these twist out to lengths of up to 68cm (27in). The sexes are also separated by colouration: the females are light fawn in colour whereas the males are dark brown to black, the darker colours making the blackbuck's white underparts and face patches more striking. Social behaviour is affected markedly by the breeding season, during which time the males, either as a small group or individually, establish a territorial defence of breeding females. Territories are marked out using urine and dung, and by rubbing scent glands against trees and bushes (the scent glands are located just in front of the eye). Breeding can take place at any time of the year, but March to May and August to October are the most intensive times. Blackbuck are currently listed as 'vulnerable' by the IUCN, their numbers having been slashed by unrestrained hunting during the nineteenth and twentieth centuries.

Species name:	*Antilope cervicapra*
Features:	Males: dark brown to black upper coat; long horns; Females: fawn coat
Habitat:	Scrub, arid lands and woodlands
Distribution:	India
Length:	Up to 1.5m (4ft 11in) without tail
Weight:	Up to 45kg (99lb)
Breeding:	One fawn after a 160- to 180-day gestation

ARTIODACTYLA: BOVIDAE

American Bison

The near destruction of the American bison (also known as the American buffalo) is an almost unparalleled natural tragedy. An estimated population of 60 million creatures around 1750 was reduced to around 750 animals by the end of the nineteenth century, the bison massacred by hunters armed with a new breed of powerful hunting rifles. Today, almost all bison (about 200,000 animals) live in protected environments, including the Yellowstone National Park in the United States and Wood Buffalo National Park in Canada. Bison are huge, solid creatures with a body length of up to 3.5m (11ft 5in) and a shoulder height of 2m (6ft 5in). They have a thick, shaggy coat concentrated at the front and mid sections of the body, with a distinctive beard on the face, while the hair around the abdomen and hind legs is short. Bison can live in climates ranging from tropical to arctic, and they rely on grasses and sedges for their main diet. Females and males tend to live separately, both within hierarchical groups, although the sexes will mix during the breeding season.

Species name:	*Bison bison*
Features:	Massive physique; shaggy coat and beard, short horns
Habitat:	Woodlands, forests, grasslands, mountains
Distribution:	Local pockets in North America
Length:	Up to 3.5m (11ft 5in) without tail
Weight:	Up to 1000kg (2200lb)
Breeding:	One calf born after a 285-day gestation

19

ARTIODACTYLA: BOVIDAE

Gaur

Native to India and Southeast Asia, the gaur is a very large wild cattle species – it can have a shoulder height of 1.8m (6ft). It was once common throughout the region, but as a result of hunting and environment destruction it is now listed as 'vulnerable' by the IUCN. Coat colours range from light brown through to black, the darkest shades being reserved for adult males. They also sport a pair of powerful upward-curving horns and have a pronounced hump at the shoulders. Gaur are found in arboreal habitats, including woodlands, monsoon forests and tropical rainforests. Here they find some degree of protection from hunters and can access the main elements of their diet – grasses, plant shoots and fruits. They tend to feed during the afternoon, but in some places they have adopted nocturnal habits to avoid daytime dangers. Gaur gather in small mixed groups of 8–40 animals, and during the rutting season the male emits strong calls that are heard up to 1.5km (1 mile) away.

Species name:	*Bos gaurus*
Features:	Hump between shoulders; dewlap under chin and forelegs; horns in both sexes
Habitat:	Tropical woodlands and forests
Distribution:	India and Southeast Asia
Length:	Up to 3.3m (11ft) without tail
Weight:	Up to 1000kg (2200lb)
Breeding:	One calf, rarely two, born after a 275-day gestation

ARTIODACTYLA: BOVIDAE

Yak

Domesticated yaks are common throughout East Asia, where they are valued for their uses in agriculture and transport, but wild yaks are endangered. The small numbers that remain after decades of hunting are found in remote, high-altitude areas of the Tibetan plateau and some adjacent areas of China and India. The wild yak survive in these hostile environments thanks to a thick undercoat of soft hair, which is covered by a heavy black or dark brown outer coat. They cope with the rarefied atmosphere thanks to a massive lung capacity and a blood system that has an improved oxygen-storage capacity over normal cattle. Grasses, herbs and lichens form the yak's diet, and it will climb to altitudes of up to 6000m (19,800ft) to find these. The yak move to lower altitudes during the winter months, but will go higher during the warmer months of August and September. Yaks are herd animals, and a typical herd consists of 30 creatures, although in exceptional circumstances this number may rise to around 100 beasts.

Species name:	*Bos grunniens*
Features:	Humped shoulders; very long outer coat
Habitat:	Treeless high-altitude terrains
Distribution:	Small pockets of South and East Asia
Length:	Up to 3.3m (11ft) without tail
Weight:	Up to 525kg (1160lb)
Breeding:	One calf born after a 258-day gestation

ARTIODACTYLA: BOVIDAE

Nilgai

The nilgai has a deep frame and sloping backline, a shape that is accentuated by the nilgai's front legs being longer than its back ones. Male colouration is grey or blue-grey, while the females have a brown or tawny coat. Males universally grow short horns of around 18cm (7in); females are occasionally known to grow horns, but this is infrequent. Nilgai are grazers and browsers in eating habits, feeding on grasses, shoots, fruit and leaves. Feeding hours tend to be confined to the periods around dawn and dusk. Its diet, and preference for easily accessible cover, means that the nilgai inhabits open areas of woodland or flat areas with scrub. It is distributed throughout South Asia, and so its main predators are tigers and other indigenous large carnivores. Nilgai herds tend to be small, around 10 animals, growing occasionally to larger groups of 50–70 creatures. Males and females remain separate until the breeding season, around September to November. Females often give birth to twin calves.

Species name:	*Boselaphus tragocamelus*
Features:	Male: short tapered horns and throat tuft; rearward–sloping profile
Habitat:	Open woodland and scrubland
Distribution:	South Asia
Length:	Up to 2.1m (6ft 10in) without tail
Weight:	Up to 300kg (660lb)
Breeding:	One or two calves born after a 240- to 258-day gestation

ARTIODACTYLA: BOVIDAE

Alpine Ibex

As its name suggests, the Alpine ibex is found in high-altitude montane pastures, living above the treeline to altitudes of around 3200m (10,200ft), but extending its range up to 6700m (22,000ft). The ibex tends to move to higher locations during the hotter summer months, dropping back to lower altitudes with the coming of winter. Its distinguishing feature is, in the male, huge sabre-like horns that can grow up to 1m (3ft 3in) in length (female horns are typically around 35cm/14in). The horns are used in battles for territorial or sexual dominance, competing males lifting themselves high on their back legs before slamming horns together with incredible force. Alpine ibex have consummate climbing skills, and they are able to negotiate precipitous terrain with sure-footed steps and a powerful jump. Males and females tend to live separately. The females will form groups of up to 20 animals, while the males will create smaller groups or live alone. The males and females will mingle, however, from late autumn through the winter for breeding.

Species name:	*Capra ibex*
Features:	Male: yellow to white patches on back and rump, long curved horns, woolly beard; females: shorter horns, lighter coat
Habitat:	Montane pasture
Distribution:	Alpine regions
Length:	Up to 1.7m (5ft 6in) without tail
Weight:	Up to 150kg (330lb)
Breeding:	One or occasionally two kids born after a 170-day gestation

ARTIODACTYLA: BOVIDAE

Mainland Serow

The serow is a goatlike animal which grows to a height of around 1m (3ft 6in) at the shoulder and has a stocky physique suited to its elevated woodland and forest habitats. Its geographical distribution reaches from the Indian Himalayas down to Thailand and Malaysia. It has a shaggy black coat and sports a relatively short pair of horns – male horns will reach a maximum length of around 25cm (11in), but they are frequently much shorter. Serows typically live in mountainous forests or woodlands up to altitudes of around 3000m (10,000ft), although they are seen on barren slopes where, unfortunately, they make easy targets for hunters. (They are often killed for their glands, which in some traditional forms of Asian medicine are reputed to have healing properties.) Serows are excellent climbers, but also capable swimmers and consequently they are found in several offshore East Asian island habitats. Their basic diet consists of leaves, shoots and some grasses. Serows tend to be solitary animals, although they meet during the October to November mating season.

Species name:	*Capricornis sumatraensis*
Features:	Flat front skull; rough hairline tracking across body; horns narrow to thin tips
Habitat:	Open montane forests or woodlands
Distribution:	Indian Himalayas down to Thailand and Malaysia
Length:	Average 1.5m (5ft) without tail
Weight:	Up to 140kg (308lb)
Breeding:	One kid born after a 230-day gestation

ARTIODACTYLA: BOVIDAE

Wildebeest

The wildebeest is one of the most recognizable of Africa's mammals, and is particularly known for its migratory habits. Once a year, the wildebeest gather into herds numbering in the thousands and, beginning in the calving season around January and February, they make huge roundtrips of up to 1600km (1000 miles). The purpose of the migration is to access better pasturelands, but the perils of crossing rivers (where many wildebeest are eaten by crocodiles) and the constant attrition from predators means hundreds of creatures are lost during the journeys. The wildebeest's main land predators are lions and hyenas, and the young are especially vulnerable. Infants are born within the middle of the herd to provide close protection, and they can stand and run after only a few minutes following birth. The infant will stay with its mother for a year, when the mother will give birth to a new calf. Male wildebeest structure themselves into small bachelor herds, establishing territories by urination, defecation and glandular secretions.

Species name:	*Connochaetes taurinus*
Features:	Silver grey and brown coat with long black mane; horns up to 80cm (32in) long in the male
Habitat:	Savannah and grasslands
Distribution:	East and southern Africa
Length:	Up to 2.4m (7ft 10in) without tail
Weight:	Up to 275kg (610lb)
Breeding:	One calf after a 250- to 260-day gestation

ARTIODACTYLA: BOVIDAE

Bontebok

The bontebok (also called the blesbok) is a striking animal with a luxurious iridescent coat that is brown with a purple sheen. A white patch dominates the front of the face, and the tail is also white. Both sexes display ringed horns, these growing to around 70cm (28in) in length. Today the creature tends to be confined to protected areas, as excessive hunting all but wiped the creature out in the nineteenth century. Preservation measures have significantly raised the numbers, but the bontebok remains on the IUCN list of endangered species, being classified as 'vulnerable'. Females and young bonteboks form themselves into herds, over which individual males attempt to establish territorial dominance. During the process of settling a territory, a male will often clash horns with other males, although rarely with enough aggression to cause serious injury. Females are able to mate from about two years old, and the infant is suckled for the first six months of life.

Species name:	*Damaliscus dorcas*
Features:	White blaze on muzzle and white underparts; rich coat colours; ringed horns in both sexes
Habitat:	Grasslands
Distribution:	Southern Africa
Length:	Up to 2.1m (6ft 10in) without tail
Weight:	Up to 155kg (340lb)
Breeding:	One calf after a 223- to 235-day gestation

ARTIODACTYLA: BOVIDAE

Thomson's Gazelle

The Thomson's gazelle can be confused with the springbok, as both adopt the stiff-legged 'pronk' run when threatened by predators. The Thomson's gazelle, however, is smaller – its average length is 0.9–1.2m (2ft 11in–3ft 11in) – and it features a thick black stripe dividing the white underparts from the fawn coat above. Its horns are also straighter and have a ringed structure almost to the very tip, and the facial markings are more elaborate than the springbok's, with black stripes on the cheeks, a dark blaze along the nose and white stripes extending from the eyes along the length of the muzzle. Socially Thomson's gazelles live in fairly large herds (around 200 creatures), which are sometimes separated by sex, but they form mixed-sex groups of thousands during annual migratory movements in search of better pasture. The female gestation period is a rapid 5½–6 months, so it can reproduce twice a year (a healthy mother is capable of conceiving only two weeks after the birth of a calf). Birth tends to take place in January to February and June to July.

Species name:	*Gazella thomsonii*
Features:	Black side stripes; long ringed horns; dark blaze on the head
Habitat:	Open grassy plains
Distribution:	East Africa
Length:	Up to 1.2m (3ft 11in) without tail
Weight:	Up to 30kg (66lb)
Breeding:	One calf born after a 150- to 180-day gestation

ARTIODACTYLA: BOVIDAE

Tahr

The tahr is an endangered species concentrated in the Himalayan regions of South Asia. Behaviourally it has much in common with the ibex, although its appearance is quite different. It has a shaggy red-brown coat in winter, the colour lightening during the summer months after the spring moult. The horns are flattened and grow to around 40cm (16in) in the males. Uncommonly, the horns can grow longer in the females. Male tahrs use their horns for typical territorial and sexual jousting; rutting males will lock their horns together and attempt to throw the other as a sign of dominance. Tahrs are shy creatures that tend to confine feeding to the early morning or dusk, hiding away among rocks and vegetation during the rest of the day. During the winter months the tahr will descend to lower altitudes to feed in the lowland forests, but in spring it will head up to heights of 5000m (16,500ft). Its diet consists of grasses and leaves. During the rutting season, the tahr will live in mixed herds of around 15–80 creatures.

Species name:	*Hemitragus jemlahicus*
Features:	Shaggy red-brown coat; short tail; flattened horns
Habitat:	Mountainous terrain and montane woodlands
Distribution:	South Asian Himalayas
Length:	Up to 1.4m (4ft 7in) without tail
Weight:	Up to 100kg (220lb)
Breeding:	One offspring after a seven-month gestation

ARTIODACTYLA: BOVIDAE

Roan Antelope

The roan antelope has a grey to red rough upper coat and white underparts, while the face has a mixed black and white colouration (the muzzle is white and there are white patches around each eye). Two distinctive features are a shaggy mane running along the ridge of the neck to the withers and a thick beard of lighter hair beneath the chin. Both males and females have these features, as well as arched horns that can grow to 1m (3ft 3in) long. These horns are used in battles for dominance among both sexes (the fights are conducted with both animals on their knees, the horns locked). A typical herd of roan antelope will consist of 12–15 females with a single dominant male. Males can group themselves in small bachelor herds of two to five animals. Their diet is mostly grass, although in times of scarcity they will also turn to leaves. They need to drink frequently – about three times a day – so they are never more than a few kilometres away from a water source.

Species name:	*Hippotragus equinus*
Features:	Long arched horns; mane along neck; light-coloured beard
Habitat:	Grasslands and sparse woodlands
Distribution:	Much of sub-Saharan Africa
Length:	2.7m (8ft 9in) without tail
Weight:	Up to 300kg (660lb)
Breeding:	One calf after a 268- to 280-day gestation

ARTIODACTYLA: BOVIDAE

Sable Antelope

Sable antelope are typically found in savannah and woodland areas, where they find the leaves, herbs and grasses that make up their diet. They are sizeable creatures: large males can grow up to 140cm (56in) at the shoulder and weigh 235kg (517lb), and their dramatic curved horns extend up to 165cm (65in). Social life is highly developed according to the sex of the creatures and the season (sable antelope herds often have separate territories for both the wet and dry seasons). The dry season tends to bring together large herds, up to and exceeding 100 creatures within a territory of around 25 sq. km (10 sq. miles), but in the wet season the groupings fragment – bachelor males form their own small groups of up to a dozen creatures. Female-only herds of 15–25 animals attract the attention of dominant territorial bulls, who share the females' territories for mating, but may spend much of their time alone. Calving is at its most intense at the end of the rainy season, after an eight- or nine-month gestation period.

Species name:	*Hippotragus nige*
Features:	White face with a dark blaze running down the centre of the muzzle; females: chestnut coat; mature males: black coat
Habitat:	Scrub, savannah and woodlands
Distribution:	East and southeast Africa
Length:	Up to 2.7m (8ft 10in) without tail
Weight:	Up to 300kg (660lb)
Breeding:	One calf after a eight- to nine-month gestation

Waterbuck

The waterbuck is a particularly large, stocky antelope that can grow up to 1.3m (4ft 4in) tall at the shoulder and weigh up to 300kg (660lb). The male's powerful-looking horns, which grow to a length of between 55 cm and 1m (1ft 9in and 3ft 3in), are its most striking feature. The rest of the body is heavyset, with short legs and a coarse red or grey-brown coat. Despite its appearance, the waterbuck is an excellent swimmer, although it tends to enter the water only as an escape route from predators. The waterbuck's social organization is highly developed. Male waterbucks tend to join bachelor herds at around 7–9 months of age, involving themselves in the frequent and violent battles for social dominance. At the age of six the males adopt territorial behaviour over areas as large as 242 hectares (600 acres). Female herds range across several male territories. The females reach sexual maturity at the age of 12–14 months, and usually give birth to one infant a year (the gestation period is about the same as that for a human being). Births tend to take place between August and November.

Species name:	*Kobus ellipsiprymnus*
Features:	Shaggy, oily red-brown or grey coat; white muzzle and patch on throat; white ring around base of tail
Habitat:	Scrub, savannah and woodlands
Distribution:	Much of sub-Saharan Africa
Length:	1.3–2.4m (4ft 3in–7ft 10in) without tail
Weight:	Up to 300kg (660lb)
Breeding:	Single calf, occasionally twins, after a 280-day gestation

ARTIODACTYLA: BOVIDAE

Kob

The slender kob antelope is an inhabitant of West, Central and East Africa. It has a graceful appearance, with a smooth golden coat and white underparts. Other distinguishing features include black stripes down the front of the legs, a white patch on the throat and white eye patches. The males grow ringed horns that are S-shaped and grow up to 69cm (28in) long. Kobs are extremely territorial animals, and the size of the male territories can be very small indeed – a male can defend a territory (known as a 'lek') only 30m (100ft) in diameter, establishing his presence with auditory signals. The lek's social behaviour is a by-product of dense herds that can number in their tens of thousands; indeed, population densities along some waterways during the dry season can reach 1000 per sq km (2500 per sq mile). However, the typical female and bachelor herds number around 5–50 creatures. Females give birth to young throughout the year, although there is a concentration of births around September to October with the end of the annual rains.

Species name:	*Kobus kob*
Features:	Black stripes down the legs; white underparts, throat bib and eye patches; S-shaped horns in the males
Habitat:	Open floodplains and grasslands
Distribution:	West, Central and East Africa
Length:	Up to 2.4m (7ft 10in) without tail
Weight:	Up to 300kg (660lb)
Breeding:	One infant after a 240-day gestation

ARTIODACTYLA: BOVIDAE

Gerenuk

The gerenuk has remarkable adaptations to survive in the harsh climate of East Africa. It is a fairly long creature, growing up to 1.6m (5ft 3in), but is extremely slender and its maximum weight is only around 52kg (115lb). Its spine is unusually flexible, and this allows it to stand on its hind legs to feed on leaves set well above the ground. Its mouth has very manipulable lips and a long tongue for stripping off the hardiest or smallest leaves from branches, and heavy thorns seem no deterrent to the gerenuk's feeding. Almost all of its fluids are obtained from its food, which includes fruit, shoots and flowers, and the gerenuk can go for indefinite periods without drinking. Gerenuks have reddish-fawn coats with white underparts and a darker band, or 'saddle', of fur running along the back and upper sides. The males also have long horns about 35cm (14in) long. Herd structure varies from mixed pairs to single-sex groupings, although the herds always tend to be small in number (rarely more than 10 creatures).

Species name:	*Litocranius walleri*
Features:	Slender neck and head; saddle of darker fur along back and sides; male has long ringed horns
Habitat:	Open grasslands and scrub areas
Distribution:	East Africa
Length:	Up to 1.6m (5ft 3in) without tail
Weight:	Up to 52kg (115lb)
Breeding:	One infant after a 210-day gestation

ARTIODACTYLA: BOVIDAE

Dik-Dik

The dik-dik is a diminutive creature growing to only 72cm (28in) and rarely weighing more than 7kg (15lb). Despite its small size, however, it is a rugged survivor. The dik-dik lives off leaves, shoots and fruit, and it can stand up on its hind legs to feed off high branches. All of the animal's water needs can be extracted from the plants, so it is not dependent upon open water for its fluid intake. Dik-diks have grey-brown coats on the upper sides with tan flanks. They are necessarily shy creatures – their small size opens them up to a wider variety of predators than larger antelope, and they are prey to jackals, hyenas, leopards and wild dogs, as well as to several large species of birds of prey. Consequently the dik-dik inhabits terrain with patches of grass, shrubs and trees for concealment, and it is nocturnal in habits. The dik-dik's social system is relatively simple and moves away from the herd system. Typically a dik-dik group will consist of a male and a female and a single offspring. They inhabit ranges of 2.5–12 hectares (6–30 acres).

Species name:	*Madoqua kirkii*
Features:	Small size; short horns on back of head
Habitat:	Arid scrub areas and grasslands
Distribution:	East and southwest Africa
Length:	Up to 72cm (28in) without tail
Weight:	Up to 7kg (15lb)
Breeding:	One calf after a 180-day gestation

ARTIODACTYLA: BOVIDAE

Rocky Mountain Goat

The Rocky Mountain goat is confined to mountainous areas of western Canada and the north and west of the United States. Its name is testimony to its superb mountaineering capabilities. Its hooves are designed to provide maximum grip on the often ice-covered terrain, the outer rim of the hooves being sharp and hard to push into crevices, while the inner section of the hoof is rubbery to provide grip. Observed climbs have seen mountain goats ascending 450 vertical metres (1500 vertical feet) in around 20 minutes. Protection from the harsh climate is provided by a thick underfur wrapped with a long yellow-white winter coat (the winter coat is moulted during the summer months). The mountain goat also has a distinctive hump of fur just atop the shoulders. The staple diet for the mountain goat is leaves, mosses, lichens, grasses and twigs. They tend to live in small groups of around four animals during the summer, although these combine into larger herds during the winter months.

Species name:	*Oreamnos americanus*
Features:	Shaggy yellow-white coat; curved horns in both sexes
Habitat:	Mountainous territories
Distribution:	Western Canada, north and western USA
Length:	Up to 1.6m (5ft 3in) without tail
Weight:	Up to 140kg (310lb)
Breeding:	One kid after a 175- to 180-day gestation

ARTIODACTYLA: BOVIDAE

Klipspringer

The klipspringer is a resident to rocky and hilly regions of eastern Africa. It is incredibly agile, with tiny hooves that provide exceptional grip on rock surfaces. Its coat is of an olive-yellow colour speckled with yellow and brown patches. The underparts of the animal and the insides of the legs are lighter than the upper coat, the colour moving to white. Male klipspringers have short, spiky ringed horns, which are little taller than the animal's large ears. Overall the klipspringer is a small creature, growing to a maximum length of 1.2m (4ft) and weighing around 18kg (40lb). It can, however, stand up on its rear legs and use all its available height to reach up to higher branches. The klipspringer diet consists of leaves, grasses, moss, flowers and shoots. All of the klipspringer's water needs can be obtained from the foliage, and even during the dry season it may not need to drink at all.

Species name:	*Oreotragus oreotragus*
Features:	Speckled olive-brown coat; short, spiky horns
Habitat:	Mountainous and rocky territories
Distribution:	Eastern Africa
Length:	Up to 1.2m (4ft) without tail
Weight:	Up to 18kg (40lb)
Breeding:	One lamb after a 214-day gestation

ARTIODACTYLA: BOVIDAE

Gemsbok

The gemsbok is a striking animal with a well-known appearance. This large creature – it can grow up to 2.4m (7ft 10in) long – has a fawn or grey body cut through with thick black stripes along the flanks and spine, terminating in a long black tail. Beneath the side stripe, the underparts are white. The legs and face are a mixed patterning, the face featuring black patches and stripes set against a white muzzle and cheeks. Both sexes sport long ringed horns up to 76cm (30in) in length, the female's being longer and thinner than the males. The gemsbok's diet is mainly grasses, shrubs, roots and tubers, melons, wild cucumbers and other available fruit, and the animal extracts most of its water from its food. Its body is superbly adapted for desert survival, losing very little water through droppings and urination. It feeds mainly at early morning and dusk (at these times, the plants are richest in dew), and it spends the middle of the day under shelter. Gemsboks breed all year round, and live in small herds of around 25 animals.

Species name:	*Oryx gazella*
Features:	Black-and-white face patterning; black side stripes; ringed horns
Habitat:	Arid grasslands and deserts
Distribution:	Southwest Africa
Length:	Up to 12.4m (7ft 10in) without tail
Weight:	Up to 210kg (460lb)
Breeding:	One calf (rarely two) after a 280- to 300-day gestation

ARTIODACTYLA: BOVIDAE

Oribi

The oribi is a delicate-looking antelope indigenous to many areas of sub-Saharan Africa. Its colouration and markings are simple: a golden fawn coat with white underparts, chin and rump. The tail is dark on the top and white underneath. On the face are white blazes around the nostrils and the eyes. Male oribis have two very short, needle-like ringed horns that angle slightly forwards at the tip. The oribi social structure is fairly loose and opportunistic, ranging from solitary creatures via male–female pairs to herds of up to 10 animals, the majority of these being females. In terms of habitat, oribis prefer open grassland, floodplains and montane grasslands, where they have good visibility to watch for predators. When threatened, oribis launch into a fast run with occasional leaps. They also make a high-pitched whistle when scared. The female oribis breed on a seasonal basis, with most of the births being concentrated in November to December, after a 210-day gestation period.

Species name:	*Ourebia ourebi*
Features:	White blazes around nostrils and eyes; two short, spiky ringed horns
Habitat:	Open grassland, floodplains and montane grasslands
Distribution:	Central and southern Africa
Length:	Up to 1.4m (4ft 7in) without tail
Weight:	Up to 21kg (46lb)
Breeding:	One lamb after a 210-day gestation

ARTIODACTYLA: BOVIDAE

Musk Ox

The musk ox is a powerful inhabitant of arctic North America and Greenland, and it is also found in Norway, Sweden and Siberia. It grows up to 2.3m (7ft 6in) long and weighs up to 440kg (900lb), and its name is derived from the powerful smell emitted by rutting males (this is generated by glands located beneath the eyes). Musk ox spend much of their lives in sub zero conditions; for insulation they have a dense, soft undercoat covered by a thick overcoat up to 61cm (24in) long, which almost touches the ground. Both males and females have downward-pointing curved horns, although the female's horns are shorter and do not have the male's prominent 'boss' where the horns meet on the head. Musk ox diet on sedges and grasses, moving from valleys to mountainsides according to the weather (winds at higher altitudes blow the snow away to expose plant growth). They have summer and winter feeding grounds, usually about 80km (50 miles) apart. Herd sizes vary between 10 and 20 animals, the herds being bigger in the winter.

Species name:	*Ovibos moschatus*
Features:	Dense, long coat; large downward-pointing horns
Habitat:	Arctic tundra
Distribution:	Extreme north of North America, Greenland and parts of far northern Eurasia
Length:	Up to 2.3m (7ft 6in) without tail
Weight:	Up to 440kg (900lb)
Breeding:	One calf after a eight- to nine-month gestation

ARTIODACTYLA: BOVIDAE

Bighorn Sheep

The bighorn sheep has several subspecies, and these live in habitats ranging from the mountains of western Canada and the United States, to desert regions of northern Mexico. They are stocky, tough creatures, protected by thick underfur and a brown outercoat of brittle guard hairs. Rams are instantly recognizable on account of their massive horns, which curve around by almost 360 degrees. These horns alone can weigh up to 13.5kg (30lb). Ewes, by contrast, have much shorter horns, with a less pronounced curl. The rams use their horns in brutal battles over females, head-butting each other for hours until one animal relents (battles lasting 24 hours have been recorded). Bighorn sheep are incredibly agile, and can sprint up near-vertical rock faces in search of food or to escape predators. Their diet consists of grasses, sedges, twigs, leaves and shoots, and they extract much of their fluid needs from their food. Bighorn sheep numbers have plummeted over the past 100 years owing to human hunting and the destruction of habitats, although current populations appear to have stabilized.

Species name:	*Ovis canadensis*
Features:	Pale patch of fur on rump; males: massive curved horns
Habitat:	Mountain slopes, foothill pastures
Distribution:	Southwest Canada, central and western USA, North Mexico
Length:	Up to 1.8m (5ft 10in) without tail
Weight:	Up to 125kg (280lb)
Breeding:	One or two lambs after a 150- to 180-day gestation

ARTIODACTYLA: BOVIDAE

Chamois

The chamois is a high-altitude inhabitant of the European Alps and the Carpathian mountains. It is a goatlike creature measuring 1.3m (4ft 6in) long in the body and weighing around 50kg (110lb). Visually it is defined by a tawny fur that darkens during the winter months, with paler fur on the underparts and white markings on the throat, jaw, cheeks and nose. Both sexes have a pair of curved horns up to 32cm (1ft) long. The chamois' principal habitat during the summer months are alpine pastures at altitudes above 1800m (6000ft), whereas in the winter it descends to forested areas roughly 1000m (3280ft) in altitude. It feeds on grasses, lichens, buds, shoots, herbs and fungi. Chamois are known for their climbing agility. Their specially designed hooves give superb grip on rock, and they can leap 2m (6ft 6in) from a standing start and jump 6m (20ft) gaps. They can also run at 50km/h (31mph) over rocky terrain. Females and young chamois live in herds of up to 30 animals, with the males tending to be solitary.

Species name:	*Rupicapra rupicapra*
Features:	Tawny fur, darkening in winter; white patches on throat and face
Habitat:	Montane pastures
Distribution:	Alpine Europe and West Asia
Length:	Up to 1.3m (4ft 6in) without tail
Weight:	Up to 50kg (110lb)
Breeding:	One kid (rarely two) after a 170- to 180-day gestation

ARTIODACTYLA: BOVIDAE

Saiga

The saiga is instantly recognizable on account of its bulging, flexible nose, which is more reminiscent of a tapir than an antelope or goat. Internally the nose has a complex structure, and its purposes may include heightening scent detection, filtering out dust and even protecting body heat in the winter (air is warmed in the nose before being passed down to the lungs). Saigas have thick coats, which are cinnamon-buff in the summer, but lighten to white in the winter. Males also have ringed horns of a waxy appearance, which grow up to 25cm (10in) long. The males use these horns during the extremely violent winter rutting, when many males are killed in combat – it has been estimated that a very harsh winter and the rutting can kill off 97 per cent of mature, sexually competitive males. Saigas are migratory animals. In the summer they form herds of around 30–40 animals, but during the autumn migrations southward and the return north in the spring tens of thousands of creatures will gather, these travelling more than 100km (60 miles) per day.

Species name:	*Saiga tatarica*
Features:	Proboscis-like nose; long semi-translucent horns
Habitat:	Grasslands and steppes
Distribution:	Central Asia
Length:	Up to 1.4m (4ft 6in) without tail
Weight:	Up to 69kg (150lb)
Breeding:	One or two calves (usually two after first year of breeding) after a 140-day gestation

ARTIODACTYLA: BOVIDAE

Bush Duiker

There are 17 different species of duiker, spread throughout Africa. The 17 species are divided into two groups of creatures: forest duikers and bush duikers. The forest duikers account for 16 of the duiker species, and as the name suggests they are adapted for life in dense forests and woods. Bush duikers, by contrast, have only one species – the common, or Grimm's, duiker – but these are the most numerous of the duikers in East Africa. Because the bush duiker inhabits open grasslands, its legs are longer for faster running and its ears are larger to provide better predator detection. The common duiker grows up to 1.2m (4ft) long, has light upper fur ranging from tawny to grey with white underparts, and features a prominent black stripe on the nose. Both sexes have very small, spiky horns. It has a broad diet, existing mainly on foliage, but also able to take fruit, berries, insects, lizards and even rodents, birds and carrion. Bush duikers lives solitary lives or as part of a pair, and the males are territorially defensive.

Species name:	*Sylvicapra grimmia*
Features:	Long, spindly legs; black nose stripe; short horns
Habitat:	Grasslands and well-watered savannah
Distribution:	Much of sub-Saharan Africa
Length:	Up to 1.2 m (3ft 11in) without tail
Weight:	Up to 25kg (55lb)
Breeding:	One fawn after a five- to seven-month gestation

ARTIODACTYLA: BOVIDAE

African Buffalo

The African buffalo, also known as the Cape buffalo, has a similarity to the Western domesticated cow, although its social structures and behaviours are more developed. It is an immensely powerful creature, the larger males reaching up to 3.4m (11ft) without the tail and weighing up to 685kg (1510lb). The buffalo is covered in a thin dark-brown coat, and both sexes have large curved horns – the male's horns meet in a prominent 'boss' in the centre of the forehead. African buffalo are intensive grazers, and they live off grasses and leaves. They cannot survive on vegetation-trapped water alone, so they stay within 15km (9 miles) of rivers and waterholes. African buffalo are gregarious animals, and can form into herds numbering more than 2000 creatures. During the dry season however, they tend to form small herds of females and young males, while mature males live in bachelor herds or become solitary. African buffalo are highly dangerous, and use sound signals to make coordinated attacks against predators and other threats.

Species name:	*Syncerus caffer*
Features:	Large curved horns meeting (in males) in a boss on the forehead; bare muzzle
Habitat:	Savannah, woodlands and wetlands
Distribution:	Much of sub-Saharan Africa
Length:	Up to 13.4m (11ft) without tail
Weight:	Up to 685kg (1510lb)
Breeding:	One calf after a 330-day gestation

ARTIODACTYLA: BOVIDAE

Eland

The eland is the world's biggest antelope species, with cowlike proportions that measure up to 3.5m (11ft) in body length and a weight of up to 1000kg (2210lb). It is greyish-fawn in colour, often with creamy vertical stripes along the flank, a black stripe running along the spine and a short mane down the back of the neck. Both males and females have twisted horns, although the male's reach up to 50cm (20in) while the female's are half that length. Eland are browsers, eating leaves, roots and tubers (it extracts underground food by digging it up with its hooves). Socially, the eland herd – which can grow to several hundred creatures – tends to consist of females and young animals; mature males group themselves into small bachelor groups (of three or four members), while older males prefer to live alone outside of breeding periods. Elands have been widely domesticated in Africa, living up to 20 years and providing milk, meat and hides, while the horns can be fashioned into tourist souvenirs.

Species name:	*Taurotragus oryx*
Features:	Large twisted horns; shoulder hump; short mane on back of neck
Habitat:	Plains and savannah
Distribution:	Central, eastern and southern Africa
Length:	Up to 3.5m (11ft) without tail
Weight:	Up to 1000kg (2210lb)
Breeding:	One calf after a 280-day gestation

ARTIODACTYLA: BOVIDAE

Four-horned Antelope

The four-horned antelope, also known as the chousingha, is so-called because of the two pairs of horns grown by adult males. Positioned just in front of the ears are the longest pair – straight, upward-pointing horns that grow up to 12cm (4.8in) long. In front of these, on the crest of the forehead, is another very short pair which grow to a maximum of 4cm (1.5in). This is a small animal growing only to 1m (3ft 3in). It has a brown coat with occasional hints of yellow and red, with pale underparts. A dark stripe runs down each leg and the muzzle, and outer ears are black. Four-horned antelope inhabit woodlands of South Asia, positioning themselves near water (they have to drink frequently) and feeding on leaves, grasses and shoots. They have an erratic style of movement, due to the fact that their rump is the highest part of their body, but they can move extremely quickly when alarmed. They tend to live on their own or in pairs.

Species name:	*Tetracerus quadricornus*
Features:	Males: two pairs of horns; dark stripes down legs; black muzzle
Habitat:	Woodland areas
Distribution:	South Asia
Length:	Up to 1m (3ft 3in) without tail
Weight:	Up to 21 kg (46 lb)
Breeding:	One calf after a 245-day gestation

ARTIODACTYLA: BOVIDAE

Nyala

The nyala is a large and visually distinctive antelope that inhabits the far southern territories of Africa. Male nyalas have a grey-brown coat with a long mane of shaggy hair from the shoulders to the rump, a strip of white hair running along the back, a series of faint white vertical stripes down the body (up to 14 in number), orange legs and a white chevron between the eyes. Large curved horns grow up to 70cm (28in). Females differ in being smaller, and in having a red-brown coat and no horns. Both sexes have a dark bushy tail. Nyala form themselves into groups numbering from two to 30, and much communication is performed with a range of unusual vocal noises. Rams, for example, bark to alert the herd to predators, and females make a clicking noise when with their young. (Newborn nyalas hide for about 18 days in long grass before joining their mother with the herd.) Nyalas like grassland habitats which have plenty of nearby water. Their diet is typical of a grazer/browser – grass, shoots, leaves, twigs, flowers and fruit.

Species name:	*Tragelaphus angasi*
Features:	Vertical white stripes; orange legs; shaggy mane; dark bushy tail
Habitat:	Bushlands close to water
Distribution:	Southern Africa
Length:	Up to 1.6m (5ft 3in) without tail
Weight:	Up to 125 kg (280 lb)
Breeding:	One infant after a 220-day gestation

ARTIODACTYLA: CAMELIDAE

Bactrian camel

The Bactrian camel inhabits small pockets of territory in inland East Asia, living mainly on the desolate steppes of the Gobi Desert. With only 1000–2000 such camels surviving in the wild, it is placed on the IUCN's endangered list, although there are significant numbers of domesticated Bactrians. The Bactrian has the ultimate adaptations for its arid environment. Its two humps store fat – these, and fluids taken from its diet of leaves and grasses, enable the Bactrian to survive for weeks on end without water. (The humps are bigger and more erect when the animal is getting plenty to eat.) When the animal finally does get to drink, it can take in 57 litres (120 pints) of water in only 5–10 minutes. Unlike the dromedary, which is suited mainly to tropical climes, the Bactrian can cope with both arctic-condition snows and equatorial heat. Its broad feet give it excellent traction on both snow and sand. Bactrians live either solitary lives or as part of packs of 6–30 females and offspring, accompanied by a single dominant male.

Species name:	*Camelus bactrianus*
Features:	Double humps; long, shaggy winter coat, shed in the summer; cleft upper lip
Habitat:	Desert and tundra
Distribution:	East Asia
Length:	Up to 3m (8ft 5in)
Weight:	Up to 690kg (1520lb)
Breeding:	One calf (rarely twins) after a 370- to 445-day gestation

ARTIODACTYLA: CAMELIDAE

Dromedary Camel

The dromedary is a now-domesticated camel that has phenomenal adaptations for survival in arid climates. Its appearance is familiar – a creature with one hump (as opposed to the two humps of the Bactrian), it stands around 2.1m (7ft) at the shoulder, with elongated legs and neck. The dromedary also has a shorter coat than the Bactrian. Its diet is hugely varied, and ranges from salty, barren plants through to animal carcasses. The hump is used to store a reserve of fat that can be broken down and consumed internally during times when food is scarce. The dromedary can survive a 40 per cent loss in body weight, double the viable loss for most animals. Camels are very resistant to dehydration – fur and fat deposits reduce sweating – and they can even drink saltwater. In desert conditions, a double row of eyelashes keep sand out of the eyes, while the nostrils are adapted to shut out blown sand and dirt. They have broad two-digit feet, which have cushioning pads ideally suited to walking on soft sand.

Species name:	*Camelus dromedarius*
Features:	Pronounced single hump (actually formed of two humps); long neck and legs
Habitat:	Arid areas
Distribution:	Northern Africa and West and South Asia
Length:	Up to 3.4m (11ft) without tail
Weight:	Up to 550kg (1210lb)
Breeding:	One calf (rarely two) after a 370- to 440-day gestation

ARTIODACTYLA: CAMELIDAE

Guanaco

The guanaco is the wild ancestor to the domesticated llama. While the domesticated animal has thrived in its protective and/or exploitative environment, the guanaco has been hunted to near destruction and is ranked as 'vulnerable' by the IUCN. (Numbers have dropped from 500 million to around 500,000 in around three centuries.) It grows up to 2.1m (7ft) in length and stands 1m (3ft 6in) at the shoulder. Gaunacos range widely in terms of habitat in their hunt for food – the diet consists mainly of grasses, shrubs, fungi and lichens. They tend to prefer cold grasslands, but will ascend to montane pastures at altitudes of up to 4000m (13,000ft). Protection from the severe climate of the Andes comes from a double coat of soft, dense underhair protected by coarse guardhair. Coat colour is various shades of brown, with white underparts and chest. The head is black with white notes around the eyes, ears and lips. A common guanaco herd consists of one male and up to seven females and young.

Species name:	*Lama guanicoe*
Features:	White underparts; black face
Habitat:	Grasslands, woodlands, arid areas, montane pastures
Distribution:	Western South America
Length:	Up to 2.1m (7ft) without tail
Weight:	Up to 130kg (290lb)
Breeding:	One calf after a 335- to 360-day gestation

ARTIODACTYLA: CAMELIDAE

Alpaca

The alpaca is, like the llama, a domesticated species derived from an extant wild species. In the case of the alpaca, its derivation is contentious – it was previously held to be a domesticated version of the guanaco, but latest research indicates that it may actually come from the vicuña (*see* separate entry). The alpaca is a mountain animal herded on the Andean grasslands of Peru, Bolivia and Chile at altitudes of 3500–5000m (12,000–16,400ft). It feeds upon the grasses indigenous to these areas, and has a robust, flexible mouth with a cleft upper lip adapted for stripping and cutting tough Andean plants. Its value to humans lies purely in its rich coat, which is used to make ponchos and other local clothing, and which is also exported widely to North American and European markets. Controlled domestic breeding has meant that the alpaca's coat is far more varied in colours than the vicuña's, and there are up to 16 different natural colour types currently listed, including shades of black, brown, white, grey and fawn.

Species name:	*Lama pacos*
Features:	Varied coat colouration; short tail
Habitat:	Grasslands, montane pastures
Distribution:	Western South America; international domestic herds
Length:	Up to 1.6m (5ft 3in) without tail
Weight:	Up to 55kg (120lb)
Breeding:	One cria after a 335- 360-day gestation

ARTIODACTYLA: CAMELIDAE

Vicuña

The vicuña is another llama-type animal from western South America. It is also the most endangered of the group, and it has teetered on the edge of possible extinction. Until the 1970s, huge numbers of vicuña were killed by hunters for their exceptional fur. However, conservation measures in Chile, Peru and Argentina have gone some way to restoring the wild numbers, and more vicuña are kept in the protective environment of zoos and safari parks. The vicuña is a delicate-looking animal (it is the smallest member of the camelid family), with a cinnamon-coloured coat with pale chest hair and lighter hair around the muzzle and throat. Its height to the shoulder ranges between 80cm (2ft 8in) and 1.1m (3ft 7in), and the creatures weigh up to 55kg (121lb). The vicuña is a classic camelid in terms of anatomy, able to survive the harsh climates of the Andes. It lives in groups of around 6–11 animals, which includes one mature male. Other males not part of a female group can form themselves into bachelor herds.

Species name:	*Vicugna vicugna*
Features:	Cinnamon coat; white chest bib
Habitat:	Grasslands, mountainous regions
Distribution:	Western South America
Length:	Up to 1.6m (5ft 3in) without tail
Weight:	Up to 55kg (120lb)
Breeding:	One infant after a 330- to 350-day gestation

ARTIODACTYLA: CERVIDAE

Moose

The moose is a herd animal occupying forest territories of northern Eurasia (where it is more commonly known as the elk) and North America. It is the largest species of deer, with a body length in males reaching 3.5m (11ft) and an adult weight of up to 700kg (1500lb). The moose's winter diet consists of twigs and leaves, particularly those from aspen, willow and poplar trees, while during the summer it also enjoys conifer shoots and aquatic plants. To acquire the latter, the moose will actually wade into rivers with its long legs and submerge its head to pull up the roots of submerged plants. Moose tend not to stray too far from water courses or swampland, because of their diet although they do undertake seasonal migrations. They often live singly or in the family group in the summer, but in the winter they cluster together into herds of around a dozen animals to afford greater protection from predators. Moose rut in September to October, and the female gestation period is around 8½ months.

Species name:	*Alces alces*
Features:	In males, pronounced dewlap (throat flap) and large antlers
Habitat:	Forested and tundra wilderness
Distribution:	Across North America and northern Eurasia
Length:	Up to 3.5m (11ft) without tail
Weight:	Up to 700kg (1500lb)
Breeding:	One or two young after a 231-day gestation

ARTIODACTYLA: CERVIDAE

Axis Deer

The axis deer, also known as the chital, is native to the grasslands and woodlands of India and Sri Lanka. They are highly attractive creatures, with luxurious red brown coats marked with pure white spots. Chitals also have white fronts to the legs, a black stripe down the spine and a white throat bib. The males have elegant branched antlers with two prongs growing upwards off the forehead, while the longer forked main antlers reach backwards, then up. Typically the antlers will grow to around 27cm (11in), and they are shed annually. In terms of social organization, axis deer are highly gregarious, living together in herds of 100-plus creatures containing a mixture of males, females and infants. A single buck is quite capable of breeding with up to 40 females. Female deer can breed throughout the year, with a gestation period of 210–238 days. Males and females can be aggressive in establishing rank, the males fighting with the antlers, while the does use their teeth and body mass as weapons.

Species name:	*Axis axis*
Features:	Red-brown coat, white spots; white throat bib; two-section antlers
Habitat:	Grasslands and open woodlands
Distribution:	India and Sri Lanka, plus imported populations in North America and Europe
Length:	Up to 1.5m (4ft 11in) without tail
Weight:	Up to 79kg (175lb)
Breeding:	One fawn after a 210- to 238-day gestation

ARTIODACTYLA: CERVIDAE

Roe Deer

The roe deer is a common resident throughout Europe and western Asia. Like many deer, it is a cautious and wary animal, a characteristic made necessary by the deer's popularity as a game animal. Typical habitats for the roe deer are forests or woodlands, environments that provide the deer with an abundance of its staple diet: grasses, shoots, herbs, berries and mushrooms. It mainly feeds in the crepuscular hours (at twilight or just before dawn), but it is active at other times. The coat of the roe deer is red-brown in the summer, moving to greyer tones in the winter. Distinctive features are a black band of fur around the muzzle and a white rump patch (the fur of the rump patch is raised when the deer is scared or alarmed). Male roe deer have spiky three-pointed antlers. Roe deer are territorial creatures, living alone or in small family groups during the summer and moving into larger herds in the winter. A mature female usually produces one or two young a year, the mother suckling the infant in a sheltered area for the first few days of the animal's life.

Species name:	*Capreolus capreolus*
Features:	Black muzzle band; white rump patch; three-pronged antlers (male)
Habitat:	Forests, woodland, grassy fields
Distribution:	Common in Europe and West Asia
Length:	Up to 1.3m (4ft 3in)
Weight:	Up to 30kg (66lb)
Breeding:	One or two young after a 294-day pregnancy

ARTIODACTYLA: CERVIDAE

Fallow Deer

The fallow deer is one of the most visually impressive European deer, with a body length of 1.3–1.6m (4ft 3in–5ft 3in). The male has large, broad antlers, and the coat colour varies from brown to black, usually with white spots along the back and rump. (Occasionally the coat will be a single colour, without the spots.) The deer form large herds of about 100 animals. Fallow deer can be seen during the day and night, but prefer the cover of twilight. Their habitats are deciduous forests, open grasslands and wooded parks. Many fallow deer are kept on private lands, the wild population having been decimated by hunting. The wild stocks have usually been deliberately introduced into the environment. Rutting season for the fallow deer is October to November, with a single infant born eight months later. It will 'lie up' in dense vegetation, remaining almost motionless until its mother returns from feeding.

Species name:	*Dama dama*
Features:	Spotted colouration on back, flanks and rump; male has broad, palmated antlers
Habitat:	Forests, woodland and grassy fields
Distribution:	Europe
Length:	Up to 1.6m (5ft 3in)
Weight:	Up to 150kg (330lb)
Breeding:	One fawn after a 230- to 240-day gestation

ARTIODACTYLA: CERVIDAE

Siberian Musk Deer

There are seven species of musk deer living throughout Asia. The Siberian musk deer – *Moschus moschiferus* – is, like many of its fellow species, endangered through hunting. Male musk deers have a musk gland located in the groin, which is used for marking territories. Unfortunately, the secretions are also used in the perfume industry, and consequently the Siberian musk deer population had dropped by 50 per cent by the beginning of the twenty-first century. The deer has a rough coat of brown to black colouration, with white striped markings on the chest and belly. In profile the deer slopes upwards from the shoulders to the hindquarters. Musk deer do not have antlers, but in the males the upper canine teeth grow extremely long – up to 10cm (4in) – and project below the animal's jawline. The deer are mainly nocturnal and live either alone or in small groups (up to three) of females and young. They live off a broad range of vegetation found at altitudes of 2600–3600m (8600–11,900ft). Musk deer are exceptionally fast runners over short distances.

Species name:	*Moschus moschiferus*
Features:	Whitish stripes on chest and belly; very short tails; males: prominent incisors extending below jawline
Habitat:	Mountainous taiga
Distribution:	Central and East Asia
Length:	Up to 1m (3ft 3in)
Weight:	Up to 18kg (40lb)
Breeding:	One to three fawns after a 200-day gestation

ARTIODACTYLA: CERVIDAE

Indian Muntjac Deer

The Indian muntjac deer, one of eight species of Asian muntjac, is common throughout South Asia, its territories stretching from northern India through Nepal and into southern China. It is a small deer, measuring up to 1m (3ft 3in) long, and is coloured in various shades of brown with pale markings. The most distinctive feature of the Indian muntjac is the bony pedicels, the bases of the animal's short antlers, which extend down the face to form prominent ridges. Male muntjac deer also grow their upper canine teeth into tusks, which are used as powerful weapons against predators and rivals. Muntjac deer inhabit forests and woodlands, where they can access a diet of grasses, leaves, shoots and fruit. They are territorial creatures and can live in small family units, the young adults being pushed out of the group with the arrival of each new fawn. However, many muntjac live solitary lives. A newborn fawn is hidden in dense undergrowth for about two to three weeks before it accompanies its mother in the outside world.

Species name:	*Muntiacus muntjac*
Features:	Prominent pedicels extending down forehead; males: tusklike upper canines
Habitat:	Woodlands and forests
Distribution:	South and East Asia
Length:	Up to 1m (3ft 3in) without tail
Weight:	Up to 20kg (44lb)
Breeding:	One fawn after a 210-day gestation

ARTIODACTYLA: CERVIDAE

White-tailed Deer

The white-tailed deer is a common deer species found dwelling in woodlands, forests and marshlands from northern South America to the southern territories of Canada, with the creatures smaller in size in the south (those in the far north are roughly the size of a red deer). It has a similar appearance to the mule deer, although it has a much longer tail. Its tail is, in fact, longer than most other deer species. The underside of the tail and the rump are white, and when the white-tailed deer senses danger or feels threatened it raises its tail and displays the white parts as a warning to others in the herd (the red deer also uses this system of display). Like most deer, the male is defined by its antlers, which begin growing at the beginning of spring. The antlers are wrapped in a velvety protective covering, which is rubbed off on trees during the summer months. The antlers then make formidable weapons during the rutting season.

Species name:	*Odocoileus virginianus*
Features:	Black muzzle; underside of body and tail white
Habitat:	Forests, woodland, grassy fields, marshlands
Distribution:	North America and northern South America
Length:	Up to 2.4m (7ft 10in) without tail
Weight:	Up to 140kg (310lb)
Breeding:	One to three fawns after a 187- to 222-day gestation

ARTIODACTYLA: CERVIDAE

Pudu

The smallest members of the deer family are the South American pudu. There are two species of pudu: the northern pudu's territory extends from Colombia down to Central Peru, while the southern pudu lives in southern Chile and western Argentina. Although the northern pudu (*Pudu mephistophiles*) is slightly bigger than the southern pudu (*Pudu puda*), both are small albeit stocky creatures. They have red-brown fur, which is more tufted at the rump, with small round ears and short legs. The ears of the northern pudu are lighter in colour than the southern pudu's, and its face and feet are dark brown to black. Pudus inhabit well-watered forests and woodlands and damp grasslands, and can be found at altitudes as high as 4000m (13,123ft). Their diet is primarily bamboo, fruit, nuts, flowers, buds and bark (rarely grass). Pudus tend to be solitary and shy creatures, confining their activities mostly to the twilight hours. The females become sexually mature within a year.

Species name:	*Pudu mephistophiles; Pudu puda*
Features:	Red-brown coat; short legs; round ears
Habitat:	Humid woodlands and forests
Distribution:	South America
Length:	Up to 85cm (34in)
Weight:	Up to 15kg (33lb)
Breeding:	One fawn after a 200- to 220-day gestation

ARTIODACTYLA: CERVIDAE

Reindeer

The reindeer, also called caribou in North America, is a common inhabitant to the far northern climes of North America and Eurasia. It is superbly adapted to life in the far north, with its thick coat providing ample insulation from the cold by trapping air, a quality that also makes the caribou a buoyant swimmer. It can lower its metabolic rate when food is scarce during the winter. Hooves are broad to suit walking on snow and mud, and their concave shape makes them ideal digging tools for foraging. The reindeer is most recognized by its large branched antlers, which are worn by both males and females – the bull's antlers can grow up to 1.2m (4ft) wide. During the rutting season, the males use their antlers in battles for mating dominance, and both sexes also use the antlers for defending against predators. Reindeer/caribou are herd animals, and in North America they make huge annual migrations, covering distances of more than 1000km (600 miles) in search of more accessible food and better calving grounds.

Species name:	*Rangifer tarandus*
Features:	North American: brown coats
	Eurasian: grey coats; black faces; broad forked antlers
Habitat:	Forests, taiga and tundra
Distribution:	Far north of North America and Eurasia
Length:	Up to 2.2m (7ft 3in)
Weight:	Up to 300kg (660lb)
Breeding:	One calf after a 227- to 229-day gestation

ARTIODACTYLA: CERVIDAE

Sambar

The sambar is one of the world's largest members of the deer family, and it can grow up to 2.5m (8ft 3in) in length. Male sambars produce magnificent forked three-pointed antlers, which grow up to 1.2m (4ft) long and are shed annually. The general appearance of the sambar is dark brown fur with redder patches on the chin, legs and tail. It also has a thick mane of fur around the neck and a black exposed muzzle. Sambar deer are a common species throughout South and Southeast Asia, inhabiting forests and woodlands. Leaves are the primary element of their diet, although they will take a variety of other vegetation. They tend to be solitary creatures, although females are seen with their young. The females will tend to breed during November and December, and give birth to one fawn after a six-month gestation period. The sambar's main predator is the tiger – a large sambar kill will keep a tiger fed for up to five days.

Species name:	*Servus unicolor*
Features:	Dark brown coat; thick mane around neck; males: large three-pointed antlers
Habitat:	Forests and woodland
Distribution:	South and Southeast Asia
Length:	Up to 2.5m (8ft 3in) without tail
Weight:	Up to 350kg (770lb)
Breeding:	One fawn after a 240-day gestation

ARTIODACTYLA: GIRAFFIDAE

Giraffe

The giraffe is a remarkable animal on many levels, not just visual. A large male specimen can tower 5m (16ft 6in) above the ground, much of the height accounted for by its long neck, which remarkably contains only seven vertebrae, like most other mammal species. It uses its height, which is accentuated by front legs longer than rear legs, to reach high branches to eat leaves, shoots and fruit. Its exceptionally long, flexible tongue and comblike teeth are ideally adapted to rake off leaves from branches. Giraffes tend to feed in the cooler hours of the day. Enormous blood pressure is required to maintain blood flow to the head, so when they drink, splaying their legs and dropping their heads low, special arterial valves reduce the flow of blood to the head to prevent possible brain damage. Giraffes form flexible herds and are non-territorial, although males will achieve dominance by 'necking' – hitting necks together in a ritualistic, dancelike action. Female giraffes give birth to one calf (rarely two) after a 15-month gestation – the newborn calf can be 2m (6ft 6in) tall.

Species name:	*Giraffa camelopardalis*
Features:	Brown-and-white patched skin pattern; 2–4 horns (ossicones); mane extending down back of neck
Habitat:	Grasslands and savannah
Distribution:	Many areas of sub-Saharan Africa
Length:	Up to 5m (16ft 6in) without tail
Weight:	Up to 1900kg (4180lb)
Breeding:	One calf (rarely two) after a 457-day gestation

ARTIODACTYLA: GIRAFFIDAE

Okapi

The okapi is the other member of the *Giraffidae* family. It is less well known than the giraffe, mainly because it hides away in the tropical rainforests of Central Africa, rather than expose itself on the sub-Saharan savannah. Indeed, it was not until 1901 that the okapi was classified as a species. The okapi is smaller than the giraffe – it reaches up to 2.2m (7ft 3in) in length – but it shares other characteristics with its larger relative. Its mouth structure is the same as a giraffe's, its diet similar, and gestation period roughly the same, while the males fight for dominance using the same 'necking' ritual. Males also share the giraffe's horn structures (ossicones). Visually, however, it is very different, with a dark brown coat (although hints of many other colours are present), white stripes on the haunches and upper legs, and a creamy white face with black muzzle. Okapis live almost entirely solitary lives, with the males establishing territories through which female okapis move.

Species name:	*Okapia johnstoni*
Features:	White stripes on legs; white lower legs with black patches; short ossicones
Habitat:	Tropical forests
Distribution:	Central Africa
Length:	Up to 2.2m (7ft 3in) without tail
Weight:	Up to 350kg (770lb)
Breeding:	One calf after a 427- to 491-day gestation

ARTIODACTYLA: HIPPOPOTAMIDAE

Pygmy Hippopotamus

The pygmy hippopotamus, as its name suggests, is a smaller version of the common hippopotamus. Its body length reaches to around 1.6m (5ft), with the tail adding another 15cm (6in), while its weight can be 275kg (610lb). Pygmy hippos also have a sleeker, squatter shape and a narrower head, while their toes are less webbed. This makes them more suited to the greater amount of time they spend on the land, where they forage for a wide variety of vegetation foods, mainly grasses, roots and fruit. Feeding tends to take place during the night, when there is less danger from predators (people are the biggest threat to the pygmy hippo). During the daytime, they tend to return to aquatic habitats, sheltering from the heat of western Africa in dens dug into the side of the river bank. Pygmy hippos mainly live alone or form small groups of no more than three animals. Their numbers in the wild have been slashed by the bushmeat trade, and they are currently classified as 'vulnerable' by the IUCN.

Species name:	*Hexaprotodon liberiensis*
Features:	Black hide; squat head
Habitat:	Forests, woodlands and aquatic environments
Distribution:	Western Africa
Length:	Up to 1.6m (5ft) without tail
Weight:	Up to 275kg (610lb)
Breeding:	One calf after a 200-day gestation

ARTIODACTYLA: HIPPOPOTAMIDAE

Hippopotamus

The hippopotamus is a remarkable creature common to Africa's rivers and lakes. Growing up to 2.7m (9ft) long and weighing 1.5 tons (1.7 tonnes), the hippo is nonetheless a graceful swimmer, powered by its stocky legs and webbed toes, and it can walk along the river bottom by expelling enough air to allow it to sink. The ears and eyes, as well as the nostrils (which are shut when the hippo is underwater), are positioned on the top of the head to allow it to keep most of its body submerged while breathing and monitoring its surroundings. Its thick skin contains mucus glands to keep the skin moist, although the hippo must not spend too long on land, or the skin will crack. Hippos tend to go to land purely for feeding; they mainly diet on grasses. In terms of social organization, female and young hippos can form groups of up to 100 individuals, although 10–20 animals is more common. Males are violently territorial, and will conduct dangerous fights with other males using their large canine teeth.

Species name:	*Hippopotamus amphibius*
Features:	Massive, broad head; short legs
Habitat:	Ponds, rivers and adjacent grasslands
Distribution:	Common throughout Africa
Length:	Up to 2.7m (9ft) without tail
Weight:	Up to 1.4 tons (1.5 tonnes)
Breeding:	One calf after a 240-day gestation

ARTIODACTYLA: SUIDAE

Babirusa

The babirusa inhabits three small islands in Southeast Asia – Sulawesi, Togian and Mangole. Male babirusas are instantly recognizable for the two long upper tusks that grow up through the snout and curve backwards toward the face. These tusks grow up to 30cm (1ft), and are flexible and brittle, and therefore frequently broken off. The babirusa's lower canines also grow long enough to project out of the mouth. Males sharpen their tusks by rubbing them on trees. Babirusas' piglike bodies are smooth and almost hairless, coloured grey to brown, and are supported by fairly spindly-looking legs. Adult babirusas can grow up to 1.1m (6ft 3in) in length and stand 80cm (2ft 7in) at the shoulder. Socially, male babirusa tend to live on their own, while females band together in groups of around eight animals. They live by foraging for nuts, leaves and fungi, and they will also eat insect larvae. The females produce one to three offspring after a five-month gestation period, the piglets being weaned slowly over six to eight months.

Species name:	*Babyrousa babyrousa*
Features:	Projecting upper tusks and lower canines (males); hairless hide
Habitat:	Tropical forests and wetlands
Distribution:	Sulawesi, Togian and Mangole
Length:	Up to 1.1m (6ft 3in) without tail
Weight:	Up to 100kg (220lb)
Breeding:	One to three piglets after a 150- to 157-day gestation

ARTIODACTYLA: SUIDAE

Giant Forest Hog

The giant forest hog is a fearsome creature, growing up to 2.1m (7ft) long and weighing up to 275kg (610lb). It is covered in coarse hair that turns from brown to black with maturity, and has a broad snout, from which project the upper canine-like tusks (these can grow to a length of 34cm/14in). Males, which are larger than females, also have bald 'warts' on the face, one under each eye. Giant forest hogs are grazing animals, avoiding the rooting activities of other pig species in preference for grasses and low-lying vegetation. They tend to feed in the early morning and late afternoon, retiring to forest or woodland nests to sleep. The hogs form themselves into mixed groups ('sounders') of up to 20 animals, with around 25 per cent of these being mature males. Battles for dominance among the males are extremely violent, involving head butts delivered at speed from distances of up to 30m (96ft). Both females and males will fearlessly tackle predators, particularly if there are young within the group.

Species name:	*Hylochoerus meinertzhageni*
Features:	Projecting canines; hairy, powerful body; tufted tail
Habitat:	Forests and woodlands
Distribution:	West, Central and East Africa
Length:	Up to 2.1m (7ft) without tail
Weight:	Up to 275kg (610lb)
Breeding:	Average two to six piglets after a 149- to 154-day gestation

ARTIODACTYLA: SUIDAE

Warthog

The warthog is a common sight throughout sub-Saharan Africa. It has a grey-brown body, which is almost hairless apart from a mane of hair running from the top of the head to the lower back. A thin, whippy tail 25–50cm (10–20in) long projects outwards from the rump, and this is held out straight when the warthog runs; a powerful adult can achieve run speeds of up to 48km/h (30mph). On the warthog's face are two visible warts beneath the eye, these being more prominent in the male, and there are two sets of tusks – the upper tusks are longer – which are potentially lethal in battles for dominance with other males and when combating predators such as lions and hyenas. Warthogs lives in sounders of up to 16 animals, and the group constructs underground burrows (or uses abandoned aardvark dens), which are lined with grasses and used for shelters. They are grazing animals, and they drop onto their front knees to eat grasses (they also feed on fruit, shoots and bulbs).

Species name:	*Phacochoerus africanus*
Features:	Double tusks; prominent mane running down spine; facial warts
Habitat:	Grasslands and savannah
Distribution:	Sub-Saharan Africa
Length:	Up to 1.5m (5ft) without tail
Weight:	Up to 150kg (330lb)
Breeding:	Typically three to four piglets after a 175-day gestation

ARTIODACTYLA: SUIDAE

African Bushpig

The African bushpig has much to distinguish it from other pig species. Its coat can be a bright rust-red colour and it has very long ears terminating in tufts. It also has white stripes running around the eyes and along the lower face. The tusks emerging from the animal's mouth wear against each other to razor sharpness, and provide potent defensive weapons. Bushpigs are primarily nocturnal animals, emerging from their burrows at night to travel along well-known trails to find food. The bushpig's diet is more varied than many pigs; not only will it take in fruit, roots and vegetation, but also insects, eggs, lizards and other small reptiles and carrion. They are also known to eat agricultural crops. Bushpigs live in areas of dense undergrowth, which are ideal for siting their secretive burrows. They group themselves into family units of four to eight animals, headed by a dominant boar. The young will stay with the family until they are about six months old, at which point the mature adults will drive them out to fend for themselves.

Species name:	*Potamochoerus porcus*
Features:	Sharp tusks; reddish coat; long, tufted ears
Habitat:	Woodlands and forests
Distribution:	West and Central Africa
Length:	Up to 1.5m (5ft) without tail
Weight:	Up to 130kg (290lb)
Breeding:	Three to eight piglets after a 125-day gestation

ARTIODACTYLA: SUIDAE

Wild Boar

Despite being hunted to near extermination in many parts of Europe during the nineteenth century, the wild boar is today common throughout Europe and Asia, and is also found in northern territories of Africa. However, its extremely cautious nature can make it a rare sighting – the creatures tend to stay hidden in shelters throughout the day, moving around at dusk and during the night, using smell and hearing for navigation. Wild boar like to live in dense undergrowth, but can also be found in woodland clearings and around marshy areas. The females and young wild boar form themselves into herds (known as 'sounders'), whereas the males tend to live solitary lives. They are powerful creatures, respected by hunters, and can run at speed. They are also good swimmers. Wild boars eat almost anything, from agricultural crops to fish. An adult female can produce a litter of four to six piglets once or even twice a year. The piglets are striped, which improves camouflage during the vulnerable early weeks of existence.

Species name:	*Suc scrofa*
Features:	Coarse grey-brown coat; mane of hair along spine; large curved tusks
Habitat:	Forests, woodland, marshlands
Distribution:	Europe, Asia, North Africa
Length:	Up to 1.8m (5ft 10in)
Weight:	Up to 200kg (440lb)
Breeding:	Four to six piglets after a 112- to 115-day gestation

ARTIODACTYLA: TAYASSUIDAE

Collared Peccary

The collared peccary is also known as the javelina, and it inhabits much of South America, extending up to parts of the Southwest United States. It is thoroughly piglike in appearance, the young creatures having a reddish coat that matures to grey, and adults have a pale collar across the shoulders. Their front feet have four-hooved toes, whereas the hind feet have only three hooves. Straight, razor-sharp canines are the peccary's defence, and it can wield these with unswerving ferocity when threatened (it will rarely attack anything unless directly threatened). The peccary is omnivorous, eating foods ranging from berries and bulbs through to small snakes and lizards, although herbivorous behaviour is preferred. They form mixed sex and age groups of up to 15 animals, which establish home territories of up to 225 hectares (555 acres). Territorial boundaries are scent-marked using musk glands located on the peccary's rump area. Peccaries create nest sites in dense thickets of vegetation or in rocky caverns. They breed throughout the year, producing one to five offspring.

Species name:	*Pecari tajacu*
Features:	Coarse grey-brown coat; piglike snout; sharp, straight tusks
Habitat:	Forests, woodland and scrub
Distribution:	South America and Southwest USA
Length:	Up to 1m (3ft 3in)
Weight:	Up to 30kg (66lb)
Breeding:	One to five piglets after a 141- to 151-day gestation

CARNIVORA: CANIDAE

Arctic Fox

The Arctic fox is superbly designed to survive in some of the most hostile environments on earth. It is covered in a luxurious thick fur, the only part left exposed being its nose. The fur changes its colour scheme according to the season. During the summer months, it is grey-brown. In the winter, the coat and underfur become up to 100 per cent thicker and turn pure white to provide superb camouflage against snowy backdrops. The Arctic fox's legs, tail, ears and muzzle are also short, to reduce heat loss from the extremities. Arctic foxes have to be extremely wily to survive the food shortages of the far northern winters. They will store food underground during the summer, and this is preserved through the winter by the permafrost. They eat almost anything, but thrive on lemmings, eggs, birds, fish and carrion, and they will often trail polar bears to eat the bear's leftovers. An average of six to 12 cubs are born in the summer, the female nursing the young for the season while the male supplies the family with food.

Species name:	*Alopex lagopus*
Features:	White coat in winter, grey-brown coat in summer; short ears and tail
Habitat:	Coastal areas and polar regions
Distribution:	Extreme north of northern hemisphere
Length:	Up to 55cm (1ft 10in) without tail
Weight:	Up to 4kg (8lb 13oz)
Breeding:	Six–12 cubs born after 60-day gestation

CARNIVORA: CANIDAE

Side-striped Jackal

The side-striped jackal is so named on account of the black and white stripes that run along the animal's flanks, although these are often indistinct and, when seen from a distance, blend in with the rest of the fur. Other features include a bushy white-tipped tail. Side-striped jackals are smaller than many other jackal types, growing to a maximum body length of around 81cm (32in). Like the blue-backed jackal, however, they form male–female pairs with offspring as their basic social unit. The pack establishes a fairly large territory, in which the adults roam and catch or scrounge a broad diet, including fruit, berries, eggs, birds, lizards, rodents and small mammals (rats, rabbits and hares are excellent catches). Nocturnal activity is preferred. The female will create an underground den for her offspring, and she will have a litter of three to five cubs after a 57 to 70-day gestation. Cubs are weaned at around 8–10 months, and leave the parents when they approach one year old.

Species name:	*Canis adustus*
Features:	White and black stripes along flanks
Habitat:	Woodlands and grasslands; forage around urban areas
Distribution:	Much of sub-Saharan Africa
Length:	Up to 81cm (32in) without tail
Weight:	Up to 14kg (31lb)
Breeding:	Three to five cubs after a 57- to 70-day gestation

CARNIVORA: CANIDAE

Golden Jackal

The golden jackal is a common and geographically widespread jackal species known for its intense family groupings. Golden jackal packs are based on a monogamous mother and father, and clusters of offspring (the packs can number up to 20 animals). The more mature offspring assist their parents in the raising of small cubs by guarding the den, grooming and collecting food. Jackal packs are extremely territorial, and will mark out an area of around 2.2 sq. km (1 sq. mile) of land with urine and the emissions from scent glands located in the face, anal and genital regions. Golden jackals are omnivorous, with carrion, small mammals, insects and fruit forming a large part of the diet. The females have a short gestation period of 63 days, producing a litter of one to nine pups. These begin to eat regurgitated food at about one month and are weaned at four months. Golden jackal coats are grey-yellow to golden, with greyer tones concentrated on the back and ginger notes on the underparts and face.

Species name:	*Canis aureus*
Features:	Grey–golden coat; bushy tail; ginger fur on belly, legs and face
Habitat:	Grasslands, arid areas, montane pastures and woodlands
Distribution:	Southeast Europe, southern Asia, North Africa
Length:	Up to 1.1m (3ft 7in) without tail
Weight:	Up to 15kg (33lb)
Breeding:	One to nine cubs after a 60- to 63-day gestation

CARNIVORA: CANIDAE

Coyote

Coyotes are found in almost every conceivable terrain within North America, from grasslands and mountains to tundra and urban environments. Their coat colour varies according to local environment, with desert coyotes having tan or grey fur, while those in mountainous or northern climes have lighter grey coats with near-white underparts. In appearance they are similar to a medium-sized domestic dog. Coyote social organization is flexible. They are frequently solitary animals, but can form male/female pairs and even larger groups when big prey needs to be hunted. They can sprint at speeds of up to 65km/h (40mph) or hunt small prey using a standing leap and vertical drop, the front legs being applied to strike and kill the animal. A mix of barks, yelps and howls enable communication with other pack members. Coyotes are most vulnerable as pups, and only 5–20 per cent of a 6–18 pup litter will survive to adulthood.

Species name:	*Canis latrans*
Features:	Grey to whitish coat depending on habitat and species; bushy tail
Habitat:	All environments types within North America except extreme Arctic
Distribution:	Most of North America
Length:	Up to 1m (3ft 3in) without tail
Weight:	Up to 16kg (35lb)
Breeding:	Six to 18 pups (typically six) after a 63-day gestation

CARNIVORA: CANIDAE

Grey Wolf

The grey wolf currently has a 'vulnerable' classification by the IUCN, its once huge global numbers reduced by centuries of hunting and persecution at the hands of humans. It is a true pack animal, living in highly organized social groups of up to 30 animals (8–12 or more common) with a complex and shifting system of rank dominance (a dominant male will head the pack). As a unit the packs hunt large animals such as caribou and moose, tracking the prey over great distances – a wolf can cover 40km (25 miles) in the day at a 10km/h (6mph) trot – and using coordinated stalk and sprint tactics to separate out weakened or young members of a herd. Wolves are also territorial creatures, marking out large areas of land with urine, faeces and scratches. Their most characteristic trait is their howl, which can carry 10km (6 miles) over open terrain. This is used to unite the pack for hunting or after sleeping.

Species name:	*Canis lupus*
Features:	Large stocky physique; long muzzle; thick fur; colour ranges from white to black, depending on age and habitat
Habitat:	Woodlands, forests, mountains and tundra
Distribution:	North America, Greenland and Eurasia
Length:	Up to 1.5m (5ft) without tail
Weight:	Up to 60kg (130lb)
Breeding:	Two to 11 pups after a 62-day gestation

CARNIVORA: CANIDAE

Maned Wolf

The maned wolf has an appearance that mixes wolf, fox and hyena. Standing about 90cm (3ft) at the shoulder, it has golden-red fur over much of its body, with black patches on the neck and muzzle and disproportionately long black legs. It is indigenous to Central and South America, living in isolated pockets in Argentina, Paraguay, Bolivia and Peru. Habitat destruction for agriculture within these countries has made the maned wolf an endangered creature; there are only an estimated 4500 in existence. Many are also shot on account of their liking for domestic chickens as a food source. Maned wolves are mainly nocturnal creatures who feed on small mammals, rodents, birds, insects, fruits and berries. They mate for life, and the male and female will occupy a territory, but they generally associate only during the May to June breeding season. Pups are born in August and September, the typical litter size being two to five.

Species name:	*Chrysocyon brachyurus*
Features:	Very long black legs; black crest on neck; reddish body fur
Habitat:	Grasslands
Distribution:	Central and South America
Length:	Up to 1.3m (4ft 6in) without tail
Weight:	Up to 23kg (51lb)
Breeding:	Two to five pups after a 62- to 66-day gestation

CARNIVORA: CANIDAE

Dhole

The dhole, also known as the Asian Red Dog or Asiatic wild dog, is an endangered inhabitant of South and East Asia. About the size of a medium-sized domestic dog (its maximum weight is about 20kg/44lb), the dhole typically has a dark red coat (there are many regional variations), with a black tail and white patches on the underparts, chest, ears and paws. Like wolves, dholes live and hunt together in territorial packs of around five – 12 animals, the pack system enabling the dhole to hunt hooved creatures that are much larger than itself. Axis deer form an ideal meal, but dholes will tackle sizeable ungulates such as bantengs. Pack coordination can even allow them to fight off tigers and wild boar. They are usually found in forests, but they range widely in terms of habitat, from mountain pastures to open plains. They are excellent sprinters in the hunt and can leap to vertical heights of 2.3m (7ft 6in). Dholes are active throughout the day, but prefer the twilight hours.

Species name:	*Cuon alpinus*
Features:	Red coat; white underparts, chest and paws; black tail
Habitat:	Forests and woodlands; montane pastures; grasslands
Distribution:	South, East and Southeast Asia
Length:	Up to 90cm (2ft 11in) without tail
Weight:	Up to 20kg (44lb)
Breeding:	One to 12 pups after a 63-day gestation

CARNIVORA: CANIDAE

African Wild Dog

Despite being hardy, adaptable creatures, African wild dogs are endangered, their numbers slashed by various forms of hunting, poisoning and trapping, road deaths and disease. Its Latin name *Lycaon pictus* means 'painted wolf', and they have impressive coat patterns made up from patches and stripes of brown, black, yellow and white fur, with a black muzzle and a black stripe running up the forehead. Wild dogs also have only four toes on each foot; canids usually have five toes on the front. Their large ears, keen eyesight, muscular bodies and powerful jaws make them formidable hunters within the context of a pack. Mixed-sex packs can number up to 30, governed by a dominant male and female (this is the pack's only breeding pair), and, unusually, males often outnumber females. Hunting together, the pack can bring down and kill animals as large as zebra and wildebeest. Litters of 2–19 pups (typically 10–12) are born in March to June, these becoming active members of the pack at about 12 months.

Species name:	*Lycaon pictus*
Features:	Elaborate coat markings; rounded ears; black muzzle and black forehead stripe
Habitat:	Savannahs, grasslands, woodlands
Distribution:	Throughout Africa
Length:	Up to 1.1m (3ft 7in) without tail
Weight:	Up to 36kg (79lb)
Breeding:	Typically 10 to 12 pups after a 69- to 73-day gestation

CARNIVORA: CANIDAE

Raccoon Dog

The raccoon dog has an appearance suggested by its name. It is a squat, strong animal measuring up to 60cm (23.6in), with a thick brindled black-brown-yellow coat and a distinctive face patterning consisting of large black patches around the eyes, a white muzzle and a bare black nose. Its short legs are covered with black fur. The coat thickens during the winter months, when the raccoon dog hibernates in an underground shelter, typically an abandoned fox or badger den. Raccoon dogs are omnivorous animals, consuming almost anything edible from fruit to shellfish. Their food gathering is assisted by superb climbing abilities; they can scale trees at speed and leap nimbly from branch to branch. Raccoon dogs prefer forest and woodland as habitats, but they also hunt and forage alongside waterways and seashores. Social groupings are flexible and include living as pairs or as small family groups. Studies have shown that raccoon dogs use latrine sites to keep their territories cleaned.

Species name:	*Nyctereutes procyonoides*
Features:	Brindled coat; black eyes patches and white muzzle; short, sleek black legs
Habitat:	Woodlands and forests; riverbanks and seashores
Distribution:	Europe; Central and East Asia
Length:	Up to 60cm (23.6in) without tail
Weight:	Up to 7.5kg (17lb)
Breeding:	Typically four to six infants after a 60- to 65-day gestation

CARNIVORA: CANIDAE

Bat-eared Fox

The face of the bat-eared fox is dominated by two huge ears. These act as vari-directional antennae when the fox is hunting; the fox walks while keeping its head close to the ground, and the ears pick up insects moving just beneath the surface. Once an insect is detected, the fox digs it up quickly with its powerful legs. The downside of this approach to feeding, particularly in open grasslands, is that the bat-eared fox leaves itself vulnerable to attack from big cats and birds of prey. Insects form the bulk of the bat-eared fox's diet, primarily harvester termites, dung beetles and grasshoppers. The diet is reflected in the nature of the fox's teeth – its teeth are much smaller than usual for a canid and hence its jaw can accommodate up to eight additional molars. Socially bat-eared foxes form male/female pairs, although offspring can live with the parents for over a year and even provide an extended family for the raising of a subsequent litter.

Species name:	*Otocyon megalotis*
Features:	Grey-buff coat going paler beneath; black accents on face and feet; end of tail black; very large ears (up to 12cm/5in long)
Habitat:	Grasslands and semi-desert
Distribution:	East and Southern Africa
Length:	Up to 66cm (26in) without tail
Weight:	Up to 4.5kg (10lb)
Breeding:	One to six cubs after a 60- to 75-day gestation

CARNIVORA: CANIDAE

Grey Fox

The grey fox, also known as the tree fox, has a distribution ranging from southern Canada to northern South America. True to its alternative name, the grey fox is a woodland dweller entirely at home among the trees, climbing and jumping with complete confidence. Such is its affinity for tree dwelling that it often creates a den in a tree hole more than 10m (32ft) above the ground. The grey fox's coat is primarily a brindled white, grey and black, with reddish fur concentrated around the neck, ears and legs, and some white fur on the underparts. It has a thick tail essential to maintaining its balance while climbing. Grey foxes tend to be nocturnal or crepuscular. They feed off small mammals, birds, eggs, fruit, fish and reptiles, among other foodstuffs, and often catch prey using a high leap and vertical drop. Like many canids, a male and female plus offspring form the basic social group, inhabiting a territory marked out with urine, faeces and scent.

Species name:	*Urocyon cinereoargenteus*
Features:	Speckled grey coat; red fur around neck and ears; whiter fur underneath
Habitat:	Woodlands; also common around urban areas
Distribution:	North America and northern South America
Length:	Up to 81cm (32in) without tail
Weight:	Up to 7kg (15lb)
Breeding:	Two to eight cubs after a 62-day gestation

CARNIVORA: CANIDAE

Kit Fox

The kit fox is confined to arid regions and grasslands of the western United States. At a maximum body length of around 52cm (20.5in), the kit fox is the second smallest of the canidae family after the fennec fox. Its coat varies in shade from tan to grey, with white underparts and a tail tipped with black. Kit foxes form a parents-and-offspring social group typical to many foxes, and they establish a den home in sand dunes or grass mounds. The diet consists of the small mammals and insects native to the region, although they will also eat some grasses, fruit and berries. Live food is caught out in the open and transported back to the den for consumption. Kit foxes are primarily nocturnal animals, and have superb senses. A reflective tapetum on the retina intensifies available light to give excellent night vision, while the ears are sensitive to the slightest sounds (they also serve to dissipate heat). Kit foxes are preyed on by coyotes and are frequently killed by humans.

Species name:	*Vulpes macrotis*
Features:	Tan to grey coat; white underparts, large, pointed ears; bushy black-tipped tail
Habitat:	Desert, semi-desert and grasslands
Distribution:	Western United States
Length:	Up to 52cm (20.5in) without tail
Weight:	Up to 3kg (6lb 10oz)
Breeding:	Three to six cubs after a 49- to 55-day gestation

CARNIVORA: CANIDAE

Red Fox

The red fox is one of the world's most familiar mammals, being common throughout North America, Europe, Asia and down into North Africa and Australia. Its success hinges upon its adaptability, and it can be found living in deserts and deciduous forests, on mountainsides and in cities. The red fox's coat ranges through various shades of orange and red, accentuated by a white lower muzzle, throat and chest, and often black on the ears and feet. It also has a thick bushy tail. Red foxes have an omnivorous diet and often divide their time between scavenging for carrion or refuse, and hunting for rabbits, mice, rats and insects (the latter are a large part of a red fox's menu). A dog and a vixen, with related offspring, create the essential social group. The foxes dig underground earth dens or simply find any convenient shelter hole, although males and non-breeding females will often simply curl up on the ground to sleep. Depending on the region, red foxes can have litters of up to 12 cubs, although four to eight are more typical.

Species name:	*Vulpes vulpes*
Features:	Red coat; white lower muzzle, throat and chest; bushy tail; black feet and ears
Habitat:	Almost all land environments, except polar and high altitudes
Distribution:	Throughout northern hemisphere; North Africa and Australia
Length:	Up to 90cm (2ft 11in) without tail
Weight:	Up to 11kg (24lb)
Breeding:	One to 12 pups after a 49- to 55-day gestation

CARNIVORA: CANIDAE

Fennec Fox

The fennec fox lives in some of the most arid areas of earth, dwelling in Saharan North Africa and extending out into the Middle East. It is a small carnivore, the smallest animal in the *Canidae* family, measuring up to 41cm (16in). Proportionately huge ears are its most conspicuous feature. These not only provide superb sound location equipment – the fennec can hear a large insect crawl across sand – but also allow body heat to radiate out into the atmosphere. Despite being a desert creature, the fennec has substantial fur, as most of its activity is confined to the cool night hours (the fennec also has excellent night vision). The fur extends between the pads of the feet to insulate the soles when walking on hot sand. Fur colour is typically a creamy yellow, lightening to white on the legs and underparts. Fennec foxes live in family groups of up to 10 animals, digging dens under rock piles and into the roots of bushes.

Species name:	*Vulpes zerda*
Features:	Creamy yellow coat; white underparts; very large ears
Habitat:	Desert and arid areas
Distribution:	North Africa and West Asia
Length:	Up to 41cm (16in) without tail
Weight:	Up to 1.5kg (3lb 5oz)
Breeding:	Two to five cubs after about a 50-day gestation

CARNIVORA: FELIDAE

Cheetah

The cheetah is known for its exceptional albeit short sprint. At top speed a large adult can achieve speeds of 100km/h (60mph) in bursts of 10–20 seconds, using the explosion of speed (which follows a period of stalking to close the distance) to catch prey such as gazelles, young wildebeest, impala and hares. Cheetahs have a golden-yellow coat covered in black spots, with black 'tear stripes' running from the eyes down to the muzzle, which has a lighter fur. The long tail is ringed with black. Compared to other big cats, the cheetah is fairly small and measures a maximum of 1.5m (5ft) in the body, compared to the 2.5m (8ft 3in) of the lion. Socially cheetahs are organized into two groups. Male cheetahs (usually a group of brothers) will form territorial packs of two to three animals. Female cheetahs will arrange themselves into a family unit of mother and offspring, although females are generally solitary if they have no cubs.

Species name:	*Acinonyx jubatus*
Features:	Golden-yellow coat; black spots; black-ringed tail; black face stripes
Habitat:	Semi-deserts and grasslands
Distribution:	Africa and West Asia
Length:	Up to 1.5m (5ft) without tail
Weight:	Up to 72kg (160lb)
Breeding:	Two to four cubs after a 93-day gestation

CARNIVORA: FELIDAE

Caracal

The caracal, also known as the desert lynx, is a relatively common member of the *Felidae* family found throughout Africa and extending into the western and southern parts of Asia. As its alternative name suggests, it is at home in semi-desert regions, although it can also be found in scrublands, grasslands, woodlands and on mountainsides. Typical coloration is red-brown, with white on the chin, throat and underparts. Its most distinctive features are the slim ears with long black tufts. Caracals are not large animals, so they confine their diet to manageable prey such as hares, rodents, young impala and other antelope, and birds. One hunting method involves a standing leap of up to 3m (9ft 10in) into the air to grab flying birds. They are also superb climbers, and often consume their catch up a tree to protect it from predators. Caracals are quite solitary apart from six-day periods of mating. Cubs stay with their mother for a up to a year, and are weaned at four to six months.

Species name:	*Felis caracal*
Features:	Red-brown coat; long tufted ears
Habitat:	Savannah, woodlands, hilly/mountainous territory
Distribution:	Africa and western and southern Asia
Length:	Up to 90cm (2ft 11in) without tail
Weight:	Up to 19kg (42lb)
Breeding:	One to four cubs after a 70- to 78-day gestation

CARNIVORA: FELIDAE

Puma

The puma is known by several other names, including panther, mountain lion and cougar, and is widely distributed through western and southern North America and down into Central and South America. It inhabits almost every type of environment within this territorial range, from the Rocky Mountains to tropical rainforests. Colouration is a plain brown, with occasionally some white around the muzzle and black on the tail and ears. It is a large, powerful cat that diets on farm livestock, deer and small mammals. It has a reputation for aggression towards people. In reality, though, it is unusual for humans to even spot these reclusive animals, which are usually solitary except when breeding. Pumas establish territorial 'home ranges' of 77–324 sq. km (30–125 sq. miles), and shelter in rocky holes or thick brush. Human persecution has endangered the puma, although in many areas conservation measures have stabilized the numbers.

Species name:	*Felis concolor*
Features:	Tawny coat; powerful physique
Habitat:	Woodlands, forests (including tropical), desert, grasslands, mountains
Distribution:	Western and southern North America; Central and South America
Length:	Up to 2m (6ft 6in) without tail
Weight:	Up to 105kg (230lb)
Breeding:	Two to three cubs after a 92-day gestation

CARNIVORA: FELIDAE

Geoffroy's Cat

The Geoffroy's cat is indigenous to Central and South America. It is a small, beautiful cat whose historic curse has been to have a luxurious black-spotted coat, the base colour ranging from yellow to silver depending on the region (the greyer animals tend to live in the south of the continent). Hunting for this coat means that the Geoffroy's cat is classified as 'vulnerable' by the IUCN, although numbers are picking up. The Geoffroy's cat is about the same size as its domestic equivalent. It differs from many domestic animals, however, by being a superb swimmer, enabling it to add fish and frogs to its diet of lizards, insects, rodents, rabbits and hares. Outside the mating season in early spring, the Geoffroy's cat is a solitary animal occupying a territory of around 2.5 sq. km (1 sq. mile). In terms of habitat, the Geoffroy's cat is found in grasslands, woodlands, on mountainsides and in semi-desert areas, and it prefers areas with plenty of cover.

Species name:	*Felis geoffroyi*
Features:	Yellow to grey coat covered with black spots; pale underparts; ringed tail
Habitat:	Woodlands, forests, arid scrublands, grasslands, mountains
Distribution:	Central and South America
Length:	Up to 66cm (2ft 2in) without tail
Weight:	Up to 6kg (13lb)
Breeding:	Two or three kittens after a 72- to 78-day gestation

CARNIVORA: FELIDAE

Northern Lynx

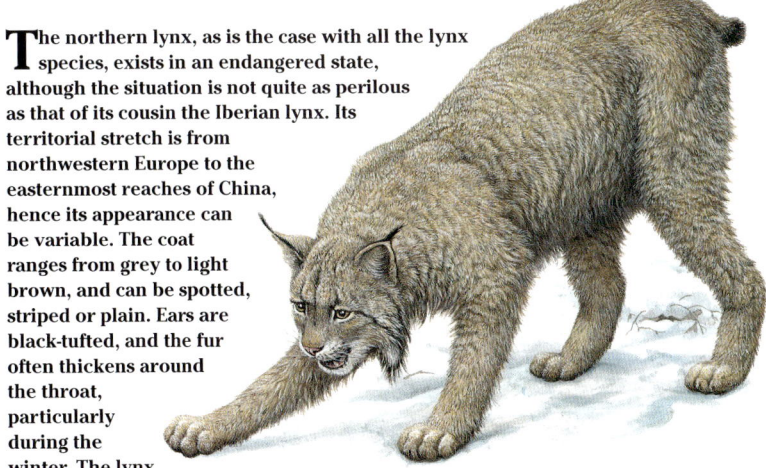

The northern lynx, as is the case with all the lynx species, exists in an endangered state, although the situation is not quite as perilous as that of its cousin the Iberian lynx. Its territorial stretch is from northwestern Europe to the easternmost reaches of China, hence its appearance can be variable. The coat ranges from grey to light brown, and can be spotted, striped or plain. Ears are black-tufted, and the fur often thickens around the throat, particularly during the winter. The lynx has large feet up to 10cm (4in) across, ideally suited for walking on deep snow. Typical habitats are woodlands, forests and mountainsides. Here the lynx hunts for rabbits, hares, deer, sheep, birds and fish. A typical home range for a lynx is about 65 sq. km (25 sq. miles). They are generally solitary, but socialize for a time when mating. The female alone is responsible for bringing up her litter of two to four kittens.

Species name:	*Felis lynx*
Features:	Variable coat patterns, spotted, striped or plain; tufted ears
Habitat:	Woodlands, forests and mountainsides
Distribution:	From northern Europe to East Asia
Length:	Up to 1.3m (4ft 4in) without tail
Weight:	Up to 38kg (84lb)
Breeding:	Two to four kittens after a 70-day gestation

CARNIVORA: FELIDAE

Sand Cat

The sand cat has a similarities to the domestic cat, but its body is specialized to cope with life in arid regions of North Africa and southwestern Asia. It is well camouflaged, with a general light-brown to grey coat colour offset by patches of white and reddish fur and black stripes. The fur is dense to provide protection from the cold desert nights, while thick hair on the soles of the feet protect the pads from hot sand and rock. Unlike many other types of cat, the sand cat is a poor climber, but it is a strong digger, using its front limbs and claws to dig up and kill gerbils, jerboas, insects and hares. It is also a snake killer, battering them to death with its paws. Its large ears are ideally suited to picking up the noise of prey both on and under the ground. Sand cats are primarily nocturnal, spending the day in a shallow den. Much about their social organization is not known, although it seems that they are generally solitary creatures.

Species name:	*Felis margarita*
Features:	Sandy-brown coat; white underparts; black stripes on legs; black tail tip
Habitat:	Arid regions
Distribution:	North Africa and arid southwestern Asia
Length:	Up to 57cm (1ft 10in) without tail
Weight:	Up to 3.5kg (7lb 11oz)
Breeding:	One to five kittens after a 60- to 69-day gestation

CARNIVORA: FELIDAE

Ocelot

Ocelot numbers plummeted during the 1960s and 1970s, when they were killed in their hundreds of thousands for their beautiful fur. The coat has a yellow/cream base on which lies a variable pattern of stripes, rosettes and spots. Ocelots live in contrasting terrains stretching from the southern United States deep into South America. They are nocturnal creatures, and during the night they hunt a wide range of prey, including rodents, small deer, rabbits, birds, small monkeys and fish. During the daytime the ocelot will generally find a tree branch on which to sleep or feed off the night's kill. Both males and females are solitary and territorial animals, using scent markings and vocal noises to mark out home ranges of between 23 and 90 sq. km (nine and 35 sq. miles); female territories are smaller. Although ocelot hunting has significantly abated, deforestation is squeezing the ocelot's habitat and putting further pressure on numbers.

Species name:	*Felis pardalis*
Features:	Yellow/cream base coat with patterns of spots, stripes and rosettes
Habitat:	Forests (including tropical rainforests and montane forests), woodlands, semi-desert and swamplands
Distribution:	Southern United States through Central and South America
Length:	Up to 1m (3ft 3in) without tail
Weight:	Up to 16kg (35lb)
Breeding:	One to two kittens after a 79- to 85-day gestation

CARNIVORA: FELIDAE

Bobcat

The bobcat is indigenous to North America from southern Canada down to Mexico. Its coat colour is generally tawny, but varies according to place, and bobcats inhabiting desert areas often have lighter coats than those in woodland or forest habitats. Many bobcats also have black spots on the coat, although the density of these varies and they may be completely absent. The bobcat's defining features are its very short black tail and its tufted ears. It is a powerful hunter and can kill game as large as deer, although it prefers rabbits, hares, gophers, squirrels and rats. Males and females both establish territories from 0.6 sq. km (0.2 sq. miles) up to 65 sq. km (25 sq. miles). The birth of young occurs almost throughout the year, with the exception of December and January, although the majority of births are concentrated in April and May. The kittens are raised by the mother only, and they will stay with her until they are around 11 months old.

Species name:	*Felis rufus*
Features:	Generally tawny coat with some black spots; very short tail; tufted ears
Habitat:	Forests, woodlands, scrubland and arid regions
Distribution:	Southern Canada through to Mexico
Length:	Up to 1.1m (3ft 7in) without tail
Weight:	Up to 15kg (33lb)
Breeding:	Two or three kittens after a 60-day gestation

CARNIVORA: FELIDAE

Serval

The serval is native to sub-Saharan Africa. It has been described as like a small cheetah – it has a yellow-tan base coat covered with black spots and stripes (the legs, neck, shoulders and tail feature stripes and rings as much as spots). Serval habitats are rarely far from water, and its dens are often found among dry reed beds and wet grasses. It is a nocturnal cat, although it can sometimes be seen during the twilight hours. Its hunting method consists of stalking its prey until close, and then pouncing on it from a distance of up to 4m (13ft 3in), the leap powered by its proportionately very long legs. The typical diet consists of rodents, hares, birds, lizards, frogs and insects, and the serval has a 50 per cent kill rate for most hunting trips. Servals are prey to leopards, African wild dogs and hyenas, and have been killed in large numbers for their fur. The fact that they control rodent populations, however, has enabled them to avoid the worst persecution from agricultural communities.

Species name:	*Felis serval*
Features:	Yellow-tan coat with black spots and stripes; long legs
Habitat:	Well-watered grasslands and wetlands
Distribution:	Much of sub-Saharan Africa
Length:	Up to 1.1m (3ft 7in) without tail
Weight:	Up to 18kg (45lb)
Breeding:	One to three kittens after a 62- to 78-day gestation

CARNIVORA: FELIDAE

European Wildcat

Wildcat species are spread throughout Europe, western and Central Asia, and Africa. The European wildcat in particular is found in territories ranging from Scotland to the Carpathians, although long periods of hunting and persecution mean that the species is still far from common. In appearance the wildcat has a grey-brown coat with black stripes running vertically down the body and horizontally around the legs. The bushy tail has black rings and a black tip. Wildcat habitats include forests and woodlands (deciduous forests are its favoured habitat), rocky coastlines, scrub and grasslands. The wildcat grows up to 75cm (30in) in body length, so its diet tends to be confined to small prey such as rats, mice, squirrels, birds, lizards, lemmings and fish. Mating takes place in February to March, with the kittens born after a 63 to 68-day gestation. The kittens are initially housed in rocky crevices, hollowed portions of trees or in abandoned animal burrows. Once they have left the mother, wildcats tend to be solitary.

Species name:	*Felis silvestris*
Features:	Grey-brown coat with black stripes; bushy black-tipped tail
Habitat:	Woodlands, forests, moorlands, scrub and coastal regions
Distribution:	Europe-wide, although sparse in many countries
Length:	Up to 75cm (30in) without tail
Weight:	Up to 8kg (18lb)
Breeding:	One to eight kittens after a 63- to 68-day gestation

CARNIVORA: FELIDAE

Margay

The margay has an appearance similar to that of an ocelot, but is generally smaller. Like the ocelot, the margay coat has a tawny to yellow base (coat colour darkens in those margays which inhabit mountainous regions) and is covered with rich black/brown spots, stripes and rosettes. The margay is unique from all other wild cats in its truly exceptional climbing capabilities. Its ankle joints can pivot through 180 degrees, and its claws are designed to dig deep into tree wood. Consequently, it runs down tree trunks head first and moves along branches while hanging upside down. It can hunt in the trees, catching prey such as small monkeys, squirrels, mice, rats, sloths, birds, possums, tree frogs and various insects. Fruit is also eaten when other food is scarce. Female margays usually have only one infant, occasionally two, and this small litter size has contributed to a dangerous decline in margay numbers, with hunting and deforestation giving the margay a 'vulnerable' classification from the IUCN.

Species name:	*Felis wiedi*
Features:	Tawny to yellow base coat with black stripes, spots and rosettes; black-ringed tail
Habitat:	Woodlands and montane forests
Distribution:	Southern North America down to northern Argentina
Length:	Up to 79cm (31in) without tail
Weight:	Up to 4kg (8lb 13oz)
Breeding:	One kitten (rarely two) after a 76- to 85-day gestation

CARNIVORA: FELIDAE

Jaguarundi

The jaguarundi has a strange appearance somewhere between that of a cat and a weasel. It typically grows between 55 and 77cm (22 and 30in) long, has a very sleek and unmarked coat that ranges from brown to red in colour, and its head appears very small in relation to its body. In terms of habitat it is found in lowland areas – typically woodlands, forests, swamplands, dry scrub and savannah – stretching from the southern United States down through South America. (The US population may have been created by the escape of jaguarundi from captivity as pets.) Jaguarundi tend to be solitary or may be found in pairs. They are diurnal and hunt various small mammals, armadillos, fish and birds. Their liking for domestic poultry brings them into direct conflict with people, although the animal's plain coat has protected them from endangerment. Litters of one to four kittens are born after the female's 70 to 75-day gestation. The infants have spotted coats, the spots disappearing with maturity.

Species name:	*Felis yagouaroundi*
Features:	Brown to red plain coat; coats tend to be darker in woodland/forest dwellers
Habitat:	Woodlands, forests, dry scrub, swamplands
Distribution:	Southern USA through South America
Length:	Up to 77cm (30in) without tail
Weight:	Up to 9kg (20lb)
Breeding:	One to four kittens after a 70- to 75-day gestation

CARNIVORA: FELIDAE

Clouded Leopard

The clouded leopard – Latin name *Neofelis nebulosa* – is so-called because of its irregular brown/black 'cloud' markings which sit on a tawny base. It also has a different bone structure in the skull from other big cats, hence it belongs to the *Neofelis* genus rather than the *Felis*. Its teeth are larger than any other big cat's proportionate to its size. In terms of overall scale, however, it is smaller than the common leopard, growing to around 1.1m (3ft 7in) in the body. While the common leopard is found from West Africa to Southeast Asia, the clouded leopard lives only in South, Southeast and East Asia. Its habitats are varied, and include tropical rainforest, swamplands, open woodlands and grasslands. Prey types include primates, porcupines, deer, birds and rodents. It hunts easily through the trees, and is an exceptional climber. Hunting and habitat loss mean that the clouded leopard is endangered, and the subspecies found on the island of Formosa may already be extinct.

Species name:	*Neofelis nebulosa*
Features:	Tawny base with striking brown/black 'cloud' markings; dark, bushy tail
Habitat:	Tropical rainforest, swamplands, open woodlands, grasslands
Distribution:	South, Southeast and East Asia
Length:	Up to 1.1m (3ft 7in) without tail
Weight:	Up to 23kg (51lb)
Breeding:	Two to five kittens after a 186- to 193-day gestation

CARNIVORA: FELIDAE

Asiatic Lion

The story of the Asiatic lion is one of terrible destruction at the hands of man. The species was once common to Eurasia from Greece to Southeast Asia, but steady persecution and hunting wiped out the animal in Europe and the Middle East by the end of World War I. Today, only around 200–300 animals survive, these being found only in the Gir forest in India. Asiatic lions are little different from African lions both physically and behaviourally. They are distinguished, however, by a fold of skin that runs along the belly, and their manes are also shorter than African lions. Asiatic lions live in the typical lion pride, but these are smaller than African prides, with around two adult females per pride (African prides have four to six females). The males tend to stay out of the pride unless feeding or mating. Sambar deer and chital are the main constituents of the Asiatic lion's diet, although it comes into conflict with people by taking goats and cattle.

Species name:	*Panthera leo persica*
Features:	Brown coat; male has tick mane; fold of skin running along belly
Habitat:	Tropical forest
Distribution:	Gir forest, India
Length:	Up to 2.8m (9ft 2in) without tail
Weight:	Up to 250kg (550lb)
Breeding:	Three or four kittens after a 100- to 119-day gestation

CARNIVORA: FELIDAE

Jaguar

The jaguar is a spectacular big cat of Central and northern South America. Although its appearance is similar to that of a leopard, it is actually more solid in the body, and its rosettes are dark within the outer rim and contain black dots or shapes. Yet like leopards, certain jaguars can be entirely black in colour, this being the result of the fur and skin containing high quantities of melanin owing to genetic variation. Jaguars are principally forest and swampland dwellers, and they enjoy environments where water is plentiful. In their range jaguars are the dominant predator (disregarding humans), and they hunt at times suited to the local environment and the movements of nearby humans. Prey for the jaguar includes deer, capybara, tapirs, fish and monkeys, and it extends to livestock such as cattle and goats. Any animal unfortunate to be caught by a jaguar is usually killed by a bite to the head from the jaguar's unusually powerful jaws (its canine teeth often penetrate the skull).

Species name:	*Panthera onca*
Features:	Brown to yellow base colour with ringed rosette patterns
Habitat:	Rainforests and swamplands
Distribution:	Central and northern South America
Length:	Up to 1.9m (6ft 3in) without tail
Weight:	Up to 160kg (350lb)
Breeding:	Two to four cubs after a 100-day gestation

CARNIVORA: FELIDAE

Leopard

Leopards are a wide-ranging species, being found in territories from West Africa to Southeast Asia. They have a familiar coat – pale yellow- to reddish-brown background with dark-rimmed, pale-centred rosettes covering the body. The tail is ringed with black, and the underparts are white. Pure black leopards – known as black panthers – are common, especially in the forest populations of Southeast Asia. Leopard size varies dramatically according to subspecies, with a weight bracket of 27–90kg (60–200lb). Leopards are consummate hunters, their carnivorous diet ranging from insects up to buffalo and giraffe. The hunt is mostly conducted by stalking the prey, then closing with a final sprint. To protect food from scavengers, the leopard often drags carcasses, even large ones, up into the branches of trees, where it can eat at leisure. The female alone cares for the litter of two cubs (average), and the cubs stay with the mother for up to two years. The mother hoists its tail high during foraging so that the cubs can easily follow her.

Species name:	*Panthera pardus*
Features:	Brown-yellow base colour with ringed rosette patterns
Habitat:	Rainforests, woodlands, savannah, montane environments; arid regions
Distribution:	Africa and across southern Asia
Length:	Up to 1.9m (6ft 3in) without tail
Weight:	Up to 90kg (200lb)
Breeding:	Two cubs (average) after a 90- to 105-day gestation

CARNIVORA: FELIDAE

Siberian Tiger

The *Panthera tigris* species is divided into five subspecies (there were previously eight, but three have recently become extinct), each subspecies being distinguished by its distribution. The Siberian tiger is the largest of the subspecies, growing up to 3m (9ft 10in) in the body and weighing up to 300kg (660lb). Few Siberian tigers remain – around 400–500 in the wild and about 1000 in worldwide zoos. Most of the wild tigers live in the Primorski Krai region of eastern Russia, with pockets in northeast China and North Korea. The Siberian tiger's coat is thicker than the Bengal tiger's, allowing it to survive in the arctic winters, and it is paler in the colour of its base and stripes. Thick fat deposits provide further insulation. Hunting mainly occurs at night when the tiger has the advantage of its superb senses (its night vision is six times stronger than a human's). Nonetheless, only about one in ten hunts is successful. Siberian tigers eat lynx, small bear, deer, wild boar, fish, rabbits and rodents, among other available foods.

Species name:	*Panthera tigris altaica*
Features:	Light orange coat with brown vertical stripes; white mane around neck; white underparts
Habitat:	Forests and woodlands
Distribution:	East Russia, northeast China, North Korea
Length:	Up to 3m (9ft 10in) without tail
Weight:	Up to 300kg (660lb)
Breeding:	Three or four cubs after a 100-day gestation

CARNIVORA: FELIDAE

Bengal Tiger

The Bengal tiger is the most numerous of the tiger subspecies – although it is still endangered – and is found in pockets in southern and eastern Asia. In India alone, numbers dropped from around 50,000 in 1900 to around 4000 today, most killed for their fur and threatened by habitat destruction. The classic coat is a luxurious orange with black vertical stripes and a white underside. The jaws and claw are massively powerful, and Bengal tigers are known to attack prey as large as young elephant and rhinos, and they have even killed crocodiles. More typical foods include chital, wild boar, monkeys, oxen and fish. Large prey is killed by clamping the throat, while the tiger attacks the back of the neck of smaller prey. Bengal tigers have a loose social structure, usually preferring solitude, but occasionally banding together into groups of three or four animals (typically a male with female and offspring). Both males and females are territorial, and the male's home range will incorporate or overlap several female territories.

Species name:	*Panthera tigris tigris*
Features:	Orange coat with black vertical stripes; white mane around neck; white underparts
Habitat:	Forests and woodlands; areas of dense undergrowth
Distribution:	Pockets throughout South and East Asia
Length:	Up to 2.8m (9ft 3in) without tail
Weight:	Up to 300kg (660lb)
Breeding:	Three or four cubs after a 100-day gestation

CARNIVORA: FELIDAE

Lion

The lion is a respected predatory carnivore native to much of sub-Saharan Africa. Typical habitat for the lion is open savannah and thin woodland – anywhere with large concentrations of ungulates. Zebra, wildebeest and antelope are the preferred kills, although a coordinated pride attack may even tackle prey as large as water buffalo. Alternatively, lions (particularly those that are old or infirm) will take smaller prey such as rabbits, hares, fish, rodents and eggs. Lions are highly social animals, and live within the structured grouping of the pride. A pride can range between two and 40 animals, the smallest prides usually consisting purely of adult males. All-male groups will attempt to take over control of a large mixed pride to win mating rights, and if successful in expelling the previous males they will stay with a pride for up to three years. The fights for pride control are extremely violent, and can easily end in death or serious injury. Males new to a pride may also kill existing cubs; this brings the bereaved mother into oestrus early, so the male can father its own children.

Species name:	*Panthera leo*
Features:	Plain tawny coat; males: thick mane surrounding neck
Habitat:	Plains and savannah, semi-desert, forests
Distribution:	Sub-Saharan Africa
Length:	Up to 2.5m (8ft 3in) without tail
Weight:	Up to 250kg (550lb)
Breeding:	One to six cubs after a 100- to 120-day gestation

CARNIVORA: FELIDAE

Snow Leopard

Although termed as a member of the leopard family, the snow leopard actually has some distinct differences from its African and southerly cousins. Its skull and vocal cord structure are unique to itself (it cannot roar), and its legs are shorter and more solid, these being better adapted to its mountainous habitat. Snow leopards live in montane environments up to an altitude of 5000m (16,500ft), preferring craggy slopes and montane forests. During the winter months, the leopard will descend to lower altitudes – this keeps it close to its prey as well as meaning it finds milder temperatures. The fur is longer and thicker than the *Panthera pardus*, and the colouration is much paler. Snow leopards are excellent stalkers, using terrain cover to close distance with wild sheep and goats, deer, marmots, hares and birds, although it can tackle larger prey such as yak. If the kill is particularly large, the leopard may consume it over a period of up to four days.

Species name:	*Panthera uncia*
Features:	Grey coat with grey-brown rosette markings; white underparts
Habitat:	Montane slopes and conifer forests
Distribution:	Central Asia
Length:	Up to 1.3m (4ft 3in) without tail
Weight:	Up to 75kg (165lb)
Breeding:	One to four cubs after a 98-day gestation

CARNIVORA: HERPESTIDAE

Meerkat

The meerkat is found in the far south of Africa, and has an appearance similar to that of the mongoose. It grows up to 35cm (14in) long in the body, with the tail adding another 17–25cm (6.7–10in). The upper fur is silver-brown, while the underparts are white and pale brown in colour. Eight bands of dark colour run across the back, and the meerkat's slender face features black rings around the eyes and white fur around the muzzle and neck. Meerkats live in colonies of up to 30 animals, utilizing abandoned rodent burrows as their home (they may even share burrows with other species). The meerkat's main food consists of insects, grubs, bird eggs, small rodents and snakes, and it has powerful front limbs and claws designed for excavating underground prey. They can also use these to dig their own burrows if others are not available, and their ears can be shut to protect them from dust and sand getting in.

Species name:	*Suricata suricatta*
Features:	Silver-brown upper fur; black eye rings; black stripes on back; black tail tip
Habitat:	Grasslands and semi-desert
Distribution:	Southern Africa
Length:	Up to 35cm (14in) without tail
Weight:	Up to 975g (1.9lb)
Breeding:	Two to four cubs after a 77-day gestation

CARNIVORA: HYAENIDAE

Spotted Hyena

Spotted hyenas, like other hyena species, are primarily scavengers, but they also form coordinated hunting packs to tackle large ungulate prey such as zebra and wildebeest. The pack will pursue the prey over distance, before attacking the animal's hind quarters and hanging on, dragging the unfortunate animal to the ground. Spotted hyenas have grey-brown coats with a thicker, darker reversed mane of hair running along the neck and back. Their name comes from the dark spots on the coat, although these fade with age. Spotted hyenas have a distinct and complex social system. It is matriarchal (the females are around 10 per cent bigger than males) and involves clans of between 30 and 80 creatures. The clan cooperates to defend its home range, which can be as large as 1000 sq. km (390 sq. miles). Clan life centres on a communal den. Apart from humans, lions are the main threat to hyenas. Lions will attack and kill clan matriarchs or other hyenas without eating them.

Species name:	*Crocuta crocuta*
Features:	Grey-brown coat with dark spots; shaggy mane along neck and back; black muzzle
Habitat:	Grasslands, savannah and semi-desert
Distribution:	Much of sub-Saharan Africa
Length:	Up to 1.33cm (4ft 6in) without tail
Weight:	Up to 70kg (155lb)
Breeding:	One to three cubs after a 100-day gestation

CARNIVORA: HYAENIDAE

Striped Hyena

The striped hyena is smaller than its brown and spotted cousins, and has a grey to light-brown coat with black stripes on the legs, parts of the torso and around the neck and lower face. It tends to live in arid mountainous areas and grasslands, and is less frequently found in desert areas or terrains where it competes for food with larger hyenas. Like other hyenas, however, the striped hyena is a scavenger, adding to carrion with a variety of live small mammals, lizards and insects and fruit. Its jaws are capable of breaking up even large ungulate bones, and the non-digestible parts of a carcass are then expelled as pellets. Hunting and foraging is conducted primarily at night, the hyena relying on its phenomenal sense of smell to locate fresh kills. Striped hyenas are typically solitary, apart from forming small, cooperative family units centred upon a communal den. They are not as doggedly territorial as the spotted hyena, but will mark out their den territory with anal-gland excretions.

Species name:	*Hyaena hyaena*
Features:	Grey to brown coat with black stripes on torso and legs; dark throat patch; bushy tail
Habitat:	Mountainous areas; grasslands and open woodland
Distribution:	Northern and eastern Africa; West and South Asia
Length:	Up to 1.1cm (3ft 3in) without tail
Weight:	Up to 40kg (88lb)
Breeding:	One to six cubs after a 88- to 92-day gestation

CARNIVORA: HYAENIDAE

Brown Hyena

The brown hyena is much less widespread than the familiar spotted hyena, its range being concentrated in southern Africa alone, mainly the Kalahari and Namib deserts. They have bushy coats coloured from dark brown to black, with a mane of contrasting lighter hair and alternating pale/dark stripes down the legs. Like most hyenas, the brown hyena has a short, black muzzle housing bone-crushing jaws. The diet is mainly carrion, although a wide variety of live food is taken, including birds and small mammals. Food is often stored for later consumption. Socially, brown hyenas live in flexible clans of 4–15 animals (the majority are females), with home ranges that extend up to 466 sq. km (180 sq. miles). Within the clan territory, the individual hyenas also control their own home range, and range marking is done through pasting a yellow excreta-like substance from anal glands. The main threats to brown hyenas come from lions, spotted hyenas and humans.

Species name:	*Parahyaena brunnea*
Features:	Thick brown to black coat with lighter mane
Habitat:	Grasslands, semi-desert, woodland savannah, scrubland
Distribution:	Southern Africa
Length:	Up to 1.3cm (4ft 6in) without tail
Weight:	Up to 47kg (105lb)
Breeding:	One to four cubs after a 90- to 95-day gestation

CARNIVORA: MUSTELIDAE

Tayra

The tayra is a common inhabitant of South America, Central America and into southern Mexico, although habitat destruction throughout this range presents long-term threats to its numbers. Its appearance is typical of a member of the weasel family, with a body length of 56–66cm (22–26in) and a tail 35–45cm (14–18in) long, which is used for balancing. The coat is made up of short, coarse grey to black fur, and the tayra also has a diamond-shaped spots on the chest ranging from yellow to red. Tayras inhabit forests and woodlands, moving across the forest floor with a flexing gait, but showing fluid movement through the trees – they are agile climbers. They hunt and forage for food mainly by day, particularly in the early morning hours, and they are omnivorous creatures – the main parts of the diet are fruit, honey, rodents, rabbits, iguanas, insects and birds. Tayras have various social units, often living alone, but sometimes forming into pairs or small groups.

Species name:	*Eira barbara*
Features:	Grey to black coat with coloured diamond throat patch
Habitat:	Forests and woodlands
Distribution:	From southern Mexico down to northern Argentina
Length:	Up to 56–66cm (22–26in) without tail
Weight:	Up to 5kg (11lb)
Breeding:	Two young after a 63- to 70-day gestation

CARNIVORA: MUSTELIDAE

Sea Otter

The sea otter has a body perfectly suited to an aquatic existence. Although it does not have the layer of blubber typically used as insulation in marine mammals, it has the thickest fur of any animal, with the hairs at a density of 150,000 per sq. cm (1 million per sq. in). Its hind feet are used to drive it through the water at speed, while the flat tail gives it excellent manoeuvrability. The front feet have retractable claws suited to gripping slippery prey. Sea otters catch much of their food underwater, and can stay underwater for up to four minutes, using their prodigious lung capacity (which is two times the size of an equivalent land mammal). The sea otter diet is typically fish, shellfish, octopus, squid and sea urchins, and they will use small rocks to smash open tough shells. They rest up in kelp forests, using wrappings of kelp to prevent them from being swept away by the currents. Sea otters live in sex-specific social groups called 'rafts'.

Species name:	*Enhydra lutris*
Features:	Thick brown coat; lighter fur on head and face; long flat tail
Habitat:	Coastal and offshore waters
Distribution:	North Pacific
Length:	Up to 1.3m (4ft 6in) without tail
Weight:	Up to 28kg (62lb)
Breeding:	One pup after a four- to nine-month pregnancy (includes two–three months of delayed implantation)

CARNIVORA: MUSTELIDAE

Wolverine

The wolverine (also known as the glutton) is a powerful carnivore indigenous to northern Eurasia and northern North America. Its body length reaches up to 1m (3ft 3in) and its appearance is bearlike – it has a shaggy black-brown coat with a pale band of fur running along the flanks and rump. A strip of pale fur also runs across the head between the ears and the eyes. Wolverine legs are stocky and the feet broad, these being well suited for travel through snowscapes. The animal is both a scavenger and a hunter. The carrion of large ungulates is an important part of its diet – its jaws are strong enough to break up frozen meat and bones – but it also hunts small deer, rodents, birds, bird's eggs and insects, as well as consuming berries and seeds. Its primary method of hunting is first to trail, then to ambush. Wolverines are diurnal animals and they are solitary; mating is the only time when mature wolverines socialize.

Species name:	*Gulo gulo*
Features:	Thick brown-black coat; pale streak down sides; white patch on chest
Habitat:	Northern forests and woodlands; mountainous terrain; tundra
Distribution:	Northern North America and northern Eurasia
Length:	Up to 1m (3ft 3in) without tail
Weight:	Up to 14kg (31lb)
Breeding:	Two or three kits after a 30- to 50-day gestation

CARNIVORA: MUSTELIDAE

Zorilla

The zorilla, also known as the striped polecat, inhabits many parts of sub-Saharan Africa. It has a skunklike appearance, although it is in fact a member of the weasel family, with black fur offset by four white stripes running from the base of the skull down the back. The fluffy tail is mixed black and white.

Like a skunk, the zorilla lifts the tail high and ejects a stinking fluid from its anal glands when it is faced with a threat (its Sudanese name means 'father of stinks'). As another way of avoiding danger, it can convincingly play dead. Zorillas avoid dense forested areas and prefer open grasslands, woodlands and scrublands. They are nocturnal, sleeping during the day in an underground burrow, but emerging at night to hunt small rodents and mammals, insects, lizards and other reptiles, frogs and birds. Pups are born after a 37-day gestation, being weaned at four to five months.

Species name:	*Ictonyx striatus*
Features:	Pure black coat with four white stripes running down back
Habitat:	Open grasslands, woodlands and scrublands
Distribution:	Sub-Saharan Africa
Length:	Up to 38cm (15in) without tail
Weight:	Up to 1.5kg (3lb 5oz)
Breeding:	One to four pups after a 37-day gestation

CARNIVORA: MUSTELIDAE

North American River Otter

The North American river otter inhabits a wide range of inland and coastal waterways, including rivers, lakes, estuaries and marshes (both saltwater and freshwater). It has all the classic aquatic characteristics seen in the Eurasian otter (*see* separate entry), although at a body length of up to 1.1m (3ft 7in) it is substantially longer than its Eurasian cousin. Typically, the otters will breed during the springtime. The embryos undergo a process of 'delayed implantation', however, their development being restricted until the gestation period is commenced in the autumn. Two to four kits are born, and home for the first month of life is a burrow set into the riverbank. North American river otters were hunted to near destruction during the nineteenth and twentieth centuries, but conservation measures introduced in the 1990s have restored the population in many areas. The main threat for the future is possibly that of industrial pollution in the animal's ecologically sensitive waterways.

Species name:	*Lontra canadensis*
Features:	Sleek red–black fur; paler fur on throat and underparts; silver whiskers
Habitat:	Inland and coastal aquatic environments
Distribution:	North America
Length:	Up to 1.1m (3ft 7in) without tail
Weight:	Up to 9kg (20lb)
Breeding:	Two to four kits after a 62-day gestation

CARNIVORA: MUSTELIDAE

Eurasian Otter

This remarkable animal is perfectly adapted for an amphibious life along the riverbanks, lake shores and coastlines of Europe and Asia. Its body length reaches 70cm (28in), but the powerful and thick tail adds another 40cm (16in). The tail, along with the webbed paws, gives the otter exceptional propulsion and manoeuvrability underwater. It has a waterproof waxy outer coat with a thick underfur to provide insulation. In terms of colouration, the otter is brown, apart from a white throat bib that extends around the lower face and beneath the eyes. Otters prefer habitats with dense, overgrown foliage on the banks or shores. It will dig a burrow (holt) directly into the bank, siting the entrance just beneath the surface of the water for added concealment. It is an exceptional hunter, using its eyesight and motion-sensitive whiskers to track down fish, frogs, eels and some small rodents and mammals on land. Eurasian otters mainly live solitary lives apart from making a male/female pairing in the spring.

Species name:	*Lutra lutra*
Features:	Brown waterproof coat; flat, angular head; webbed paws; white throat bib and lower face
Habitat:	Rivers, lakes and coastlines
Distribution:	Across Europe and Asia
Length:	Up to 70cm (28in) without tail
Weight:	Up to 10kg (22lb)
Breeding:	Two or three young after a 60- to 70-day gestation

CARNIVORA: MUSTELIDAE

Eurasian Pine Marten

Eurasian pine martens are found in deciduous, mixed and conifer forests across western Europe and into the reaches of western Siberia in the east. They grow up to 55cm (22in) in body length, and the tail adds another 26cm (10in). The coat is a luxurious brown fur, which thickens in the winter and thins out during the summer months. A distinguishing mark is a cream to orange patch of fur over the throat and chest. Pine martens establish a home territory utilizing several abandoned nests, burrows and convenient crevices or tree holes as temporary homes. Male home ranges overlap with those of females, and the animals are generally solitary apart from during the breeding season and when a mother is attendant upon infants. There is also evidence that the winter home ranges are up to 50 per cent smaller than those occupied in the summer. The pine marten diet includes mice, rats, voles, reptiles, birds, snails, insects and fruit, and food is stored during the summer and autumn in preparation for the winter.

Species name:	*Martes martes*
Features:	Brown fur; cream to orange fur on throat and chest
Habitat:	Woodlands and forests
Distribution:	Across Europe to West and North Asia
Length:	Up to 155cm (22in) without tail
Weight:	Up to 2kg (4lb 6oz)
Breeding:	Average three young after a seven to eight-month pregnancy (includes several months of delayed implantation)

CARNIVORA: MUSTELIDAE

Fisher

The fisher is a marten-like carnivore found across Canada and the northern United States. Common prey for the fisher are shrews, squirrels, chipmunks and porcupines; indeed, the fisher's preference for chipmunks saved it from complete extermination during the 1940s, as it was realized that the porcupine population would become a critical problem without fishers to control it. The fisher has a dark brown to black colouration, the fur being long, rough and shaggy. Fishers are just as comfortable up in the trees as down on the ground, and the habitat destruction resulting from logging is the fisher's biggest environmental threat (trapping is strictly controlled by the US/Canadian governments). Typically a fisher will establish a home range of 0.6–10 sq. km (1–4 sq. miles), set through scent markings, and territorial clashes between males are extremely violent. Female fishers, like many in the *Mustelidae* family, have a delayed implantation of their embryos, so the time from conception to birth can be as long as 51 weeks.

Species name:	*Martes pennanti*
Features:	Rough brown to black fur; tail and legs darker
Habitat:	Woodlands and forests
Distribution:	Across Canada and the northern USA
Length:	Up to 75cm (30in) without tail
Weight:	Up to 5kg (11lb)
Breeding:	Average three young after a 12-month pregnancy (includes up to 11 months of delayed implantation)

CARNIVORA: MUSTELIDAE

Eurasian Badger

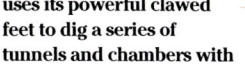

The Eurasian badger is one of the larger land members of the *Mustelidae* family, growing up to around 90cm (35in) in the body, with a tail up to 20cm (35in). It is common throughout much of Eurasia, its habitat being forests and woodlands, where it can dig a sett camouflaged by foliage. The sett is an extensive earthwork; the badger uses its powerful clawed feet to dig a series of tunnels and chambers with multiple entrances. Within the badger sett lives the typical clan of around six badgers (one male, one or two females and cubs). In appearance, the badger has a grey or grey-brown upper coat, black underparts and legs, and a white head with two black stripes, one running over each eye. Badgers are nocturnal creatures, coming out at night to gather insects, earthworms, slugs, birds, small rodents, eggs, berries, mushrooms and many other elements of an omnivorous diet. Like many mustelids, the female badger can delay implantation of an embryo by up to 10 months.

Species name:	*Meles meles*
Features:	Grey upper fur; black legs; black stripes on white face
Habitat:	Woodlands and forests
Distribution:	Across Europe and Asia
Length:	Up to 90cm (35in) without tail
Weight:	Up to 12kg (26lb)
Breeding:	Average six cubs after a 49-day gestation (following up to 10 months of delayed implantation)

CARNIVORA: MUSTELIDAE

Ratel

The ratel is also known as the honey badger. It has a powerfully developed physique and a reputation for being one of the world's toughest animals. Its skin is extremely thick and loose on the body, thereby making it difficult for predators to grasp. Ratels are known to be ferocious fighters, and have chased mature lions away from kills on occasions. They are omnivorous creatures, feeding off young or small mammals, rodents, reptiles, birds, frogs, fruit and carrion. They also eat snakes, having a high degree of immunity to snake bites. Honey badgers also have great endurance, so they can chase down prey over long distances until it collapses from exhaustion. They will also raid bees' nests, sedating the occupants with foul-smelling anal emissions before pulling the nest apart. In appearance, the ratel has a silver-grey upper body, with black or brown fur on all other parts. The ears are recessed inside the skin, and they can be shut during digging to keep out dirt.

Species name:	*Mellivora capensis*
Features:	Silver-grey upper head, back and tail; black or brown fur beneath
Habitat:	Woodlands, montane forests, savannah, semi-desert
Distribution:	Across Africa (except Sahara); West and South Asia
Length:	Up to 77cm (30in) without tail
Weight:	Up to 13kg (29lb)
Breeding:	One to four young after a 6-month gestation

CARNIVORA: MUSTELIDAE

Striped Skunk

The striped skunk is instantly recognizable – it has a black coat bisected by thick white stripes, which run from the top of the head down to the tail. It also has a thin white stripe along the muzzle. The skunk is most known for its malodorous defence when threatened. From two anal ducts the skunk can eject a foul yellow musk liquid across distances of up to 3m (9ft 9in), the fluid stinging the eyes as well as producing an overpowering stench. To fire the liquid, it raises its tail high and turns its back to face the predator; the skunk has little speed so it relies heavily on its static defence. The skunk's habitat is usually mixed grasslands, woodlands and forests, and its scavenging habits also bring it among human habitations. It diets on everything from grasses to carrion, and most of its food is collected at night. Skunks shelter throughout the day in abandoned burrows and hollows. They do not hibernate, but during the winter they can enter long periods of dormancy.

Species name:	*Mephitis mephitis*
Features:	Black fur with two white stripes running from head to tail
Habitat:	Grasslands, woodlands, forests, around human habitation
Distribution:	North America
Length:	Up to 75cm (30in) without tail
Weight:	Up to 6.5kg (14lb)
Breeding:	Six to eight infant cubs after a 63-day gestation

CARNIVORA: MUSTELIDAE

Stoat

The stoat is found in a wide range of terrains and climates throughout the northern hemisphere. It has a close visual similarity to the weasel, being of roughly the same proportions and physique. The stoat in northern snow climates develops a pure white winter coat (these are called ermine). Otherwise the stoat has white underparts and an upper coat in various shades of brown. Stoats create territories typically of 2000–4000 sq. metres (21,500–43,000 sq. ft) centred on a den, although the territories can be twice as large in open, barren countries such as Russia. The den itself is anywhere that provides shelter and concealment, with abandoned animal burrows being favoured. Stoats are committed predators, using their excellent sense of smell and astonishing endurance to track down prey and kill it with a powerful bite to the base of the skull. Stoats feed on small mammals such as rats and hares, and also eat birds and insects.

Species name:	*Mustela ermina*
Features:	Brown to ginger fur with white underparts. In Northern snow climates entirely white apart from black tail tip during winter
Habitat:	Grasslands, woodlands, forests, mountains, arctic
Distribution:	Across the northern hemisphere
Length:	Up to 24cm (9.4in) without tail
Weight:	Up to 200g (7oz)
Breeding:	Five to 12 kits after a 21- to 28-day gestation

CARNIVORA: MUSTELIDAE

European Polecat

The European polecat is indigenous to most of Europe except its extreme northern climes (in the United Kingdom the polecat is confined to Wales), and is closely related to the domestic ferret. Adult European polecats are 35–51cm (14–20in) long in the body, and they have black or dark fur coats, with lighter underfur visible beneath the hairs. In terms of habitat, the European polecat likes forests (including montane forests) and woodlands, but it is also found alongside aquatic environments such as rivers and lakes. The polecat is well adapted to the latter habitat, as it is an excellent swimmer and diver, and can pursue fish and frogs underwater. On land it feeds on small rodents, birds, eggs, worms and insects, and its tendency to massacre poultry has led to widespread trapping by farmers. A typical polecat home range is around 100 hectares (247 acres), and the animals are solitary and nocturnal in behaviour.

Species name:	*Mustela putorious*
Features:	Dark-brown to black fur with a visible cream or white underfur
Habitat:	Woodlands, riverbanks and lakesides
Distribution:	Europe
Length:	Up to 51cm (20in) without tail
Weight:	Up to 1.5kg (3lb 5oz)
Breeding:	Five to Eight kits after a 40- to 42-day gestation

CARNIVORA: MUSTELIDAE

European Mink

The European mink is little different from the American mink, except that it is smaller – the largest specimens reach a body length of around 40cm (16in), whereas the American mink can grow up to 54cm (21.3in). The fur is dark brown, blackening at the tip of the tail, and it has some white fur around the lips. The European mink's status in the wild is now critical, with isolated pockets of the animals scattered across Europe, often with great distances between them, and the IUCN has given the species an 'endangered' category. In the late 1800s, the animals were common across Europe, but a combination of factors – hunting, habitat destruction and water pollution – led to a freefall in the population. Furthermore, the American mink was introduced into Europe in 1926, and these preyed on their smaller cousins or dominated the mink's part of the food chain. Many European mink are now being bred in captivity and released into the wild, but the future of the species in the wild in uncertain.

Species name:	*Mustela lutreola*
Features:	Dark-brown fur
Habitat:	Streams, riverbanks and lakesides
Distribution:	Isolated pockets in Europe
Length:	Up to 40cm (16in) without tail
Weight:	Up to 800g (1lb 12oz)
Breeding:	Two to seven young after a 39-day gestation (following period of delayed implantation)

CARNIVORA: MUSTELIDAE

Weasel

The weasel is a slender mustelid common throughout North America and Eurasia. Like the stoat, it has dark upper fur – typically in red or brown tones – with white underparts. Its tail is somewhat shorter than the stoat's, and it does not feature the black tip seen at the end of the stoat's tail. In very cold northern climates, the coat will turn to pure white during the winter months. Hunting is a constant and voracious activity, and despite its small size (its body length is a maximum of 24cm/9.4in) it can tackle and kill prey five times its size, including rabbits, poultry and large rats. Each paw has five razor-sharp claws, and its bite is extremely potent. The weasel's most dangerous predators are birds of prey such as owls and hawks. Infant weasels are raised purely by their mothers, the father engaging the female only for mating. The young weasels will be weaned at around four weeks, and by three months they will begin to hunt for themselves.

Species name:	*Mustela nivalis*
Features:	Brown fur with white underparts; turns white in northern winters
Habitat:	Grasslands, woodlands, forests, mountains, arctic
Distribution:	Across the northern hemisphere
Length:	Up to 24cm (9.4in) without tail
Weight:	Up to 250g (9oz)
Breeding:	Three to 10 infants after a 35- to 37-day gestation

CARNIVORA: MUSTELIDAE

American Mink

Unlike the imperilled European mink, the American mink is common throughout many regions of Europe and Asia and in its North American homeland. These versatile predators are dark brown to black in colour, although some 10 per cent are grey-blue. The word 'opportunistic' is often applied to the American mink's hunting habits, and its diet includes small mammals and birds and many aquatic creatures – the mink's preferred habitat is forested areas bordering on streams, rivers and lakes. Its feet are partially webbed to aid swimming, and it can dive to depths of 5m (16ft 5in). Mink make dens in the banks of their aquatic habitats, giving the den warmth through importing grasses, leaves and the fur of kills. They are territorial and solitary, with the home range being from 1–5km (0.6–3 miles) in diameter – female territories are smaller – and marked out with musky anal secretions, urine and faeces.

Species name:	*Mustela vison*
Features:	Dark brown to black fur; occasionally grey-blue
Habitat:	Streams, riverbanks and lakesides
Distribution:	Across North America and Eurasia
Length:	Up to 54cm (21.3in) without tail
Weight:	Up to 2kg (4lb 6oz)
Breeding:	Two to seven young after a 39-day gestation (following period of delayed implantation)

CARNIVORA: MUSTELIDAE

Giant Otter

The giant otter truly lives up to its name – it can grow up to 1.4m (4ft 6in) from nose to rump, with the tail adding another 65cm (26in). Apart from the size, however, its appearance is similar to most other species of otter. It has a velvety dark-brown fur with a distinctive pattern of cream to yellow spots and stripes from the chin down to the chest. Its habitats are the tropical waterways of north and central South America; it prefers freshwater rivers and lakes, but will also inhabit reservoirs and agricultural waterways. The diet is mainly fish, of which it can eat 2.7–4kg (6–9lb) each day, but it is large enough to kill a small cayman or an anaconda if fish supplies are low. Typical of otters, the giant otter forms parental/offspring groups (known as holts) of 5–10 animals. The birthing season is May to September, and young otters will start to leave the den on hunting trips when they are around three to four months old.

Species name:	*Pteroneura brasiliensis*
Features:	Dark-brown fur; creamywhite throat and chest markings
Habitat:	Streams, riverbanks and lakesides
Distribution:	North and central South America
Length:	Up to 1.4m (4ft 6in) without tail
Weight:	Up to 32kg (71lb)
Breeding:	One to three cubs after a 65- to 70-day gestation

CARNIVORA: MUSTELIDAE

Eastern Spotted Skunk

The eastern spotted skunk is much smaller than the striped skunk, reaching a maximum body length of only 34cm (13.4in). Its colouration, however, is far more elaborate, the coat alternating between black and white stripes and other markings with a white spot centred on the forehead. The spotted skunks share much in common with other species of skunk: they have the same threat response (the spraying of a noxious liquid from the anal glands), the same diet and a similar habitat range. Dens are made in rocky cracks and crevices, or in appropriated animal burrows. They also like to make homes beneath human dwellings, particularly agricultural buildings in which they can find plenty of rats and mice to eat. They are superb climbers, and will nest in trees and even the attics of buildings. The main animal threats to the eastern spotted skunk are dogs, cats, foxes, coyotes and owls, although their smelly defence will repel most of its predators if it is deployed in time.

Species name:	*Silogale putorius*
Features:	Variegated black-and-white coat; white forehead spot
Habitat:	Forest and woodland areas, grass prairies, rocky terrain, urban
Distribution:	East and central USA; northern Mexico
Length:	Up to 34cm (13.4in) without tail
Weight:	Up to 1kg (2lb 3oz)
Breeding:	Two to nine young after a 50- to 65-day gestation

CARNIVORA: MUSTELIDAE

American Badger

The American badger has several features to distinguish it from the Eurasian badger. Its body has a more flattened appearance and its head is triangular. The fur is a brindled brown and grey, but there are white markings on the side of face and a long white stripe running from just behind the turned-up nose to the back of the head. The feet are black and there is yellow to white fur on the underparts. American badgers are found in a range from southwest Canada down to northern Mexico, and its habitats are plains and prairies, woodland (particularly that bordered by open country) and agricultural areas. The typical diet of the American badger is ground squirrels, gophers, kangaroo rats and cottontails – these are dug out of the ground using the badger's powerful clawed front feet. American badgers tend to be solitary apart from the mating season, during which time the male will have more than one mate. The young tend to be born in March and April.

Species name:	*Taxidea taxus*
Features:	Grizzled grey-brown upper coat; white stripe from nose to forehead; yellow underparts; black feet
Habitat:	Plains and prairies, woodlands, agricultural areas
Distribution:	Southwest Canada down to northern Mexico
Length:	Up to 72cm (28in) without tail
Weight:	Up to 12kg (26lb)
Breeding:	One to five young after a 42-day gestation (5½-months of delayed implantation)

CARNIVORA: PHOCIDAE

Elephant Seal

Elephant seals are separated into two species, the southern elephant seal (*Mirounga leonina*) and the northern elephant seal (*Mirounga angustirostris*). Both are enormous creatures, and the male southern elephant seal can reach up to 6m (20ft) in length and weigh an enormous 5 tons (5.5 tonnes). Males are up to five times heavier than the females. Northern elephant seals are found in the northwest Pacific region, whereas their southern relative inhabits Antarctic and South Atlantic waters. The distinguishing feature of the elephant seal is the male's enlarged 'trunk' on the nose. This is inflated during the noisy and often violent clashes with other males during the breeding season. The females will produce a single pup after a seven-month gestation following a four-month delayed implantation. Both species of elephant seal are formidable divers, and one male northern elephant seal was recorded at a depth of 1569m (5150ft). The elephant seal diet includes a wide range of marine organisms, including fish, squid, sharks and crabs.

Species name:	*Mirounga spp.*
Features:	Silver-grey fur; male has large trunklike inflatable nose
Habitat:	Coastal and offshore aquatic environments
Distribution:	Northern Pacific; Antarctic and South Atlantic waters
Length:	Up to 6m (20ft)
Weight:	Up to 5 tons (5.5 tonnes)
Breeding:	One pup after a 350-day pregnancy

CARNIVORA: PROCYONIDAE

Red Panda

The red panda, also known as the lesser panda, belongs to a different family than the giant panda, and actually looks more like a chestnut-coloured raccoon than a bear. The red-brown fur turns black on the belly and the limbs, and the face has white markings around the muzzle and eyes, and along the ears. Dark and light rings of fur alternate along the longer bushy tail. Habitat loss has caused a plummet in the number of wild red pandas, the remainder concentrated in a pocket of South and Southeast Asia. Its typical habitat is mountainous forest slopes, where the red panda find its diet of bamboo leaves and shoots, acorns, berries, fruit, roots, insects, birds' eggs (and chicks) and occasionally small mammals such as mice. Red pandas are nocturnal creatures and live solitary lives, unless they are mothers with cubs. While slow on the ground, their retractable claws make them agile climbers, and females will nest in vegetation-lined tree holes to give birth.

Species name:	*Ailurus fulgens*
Features:	Red-brown fur; ringed tail; black legs and belly; white facial markings
Habitat:	Forests and woodlands, particularly on mountainous slopes
Distribution:	South and Southeast Asia
Length:	Up to 64cm (25in) without tail
Weight:	Up to 6kg (13lb)
Breeding:	One to four young after a 90-day gestation

CARNIVORA: PROCYONIDAE

Kinkajou

The kinajou is a relative of the red panda, although the appearance is quite different. It has a uniform thick brown fur, which has a dense, woolly quality, and some lightening of the colour around the small, rounded face. Kinkajous have very long, thick prehensile tails extending up to 57cm (22.4in), and these are used for balance and for wrapping around branches when climbing. They are found in tropical forests from southern Mexico to southern Brazil, and they rarely need to venture down to the forest floor. Although kinajous are carnivores, and will eat insects, grubs and even small mammals, the main part of the diet is fruit, flowers, honey and nectar; the kinkajou has a very long tongue, up to 12.7cm (5in), and this is used to access the last two items on this list. They are nocturnal creatures, spending most of the day asleep in tree hollows, which is also where the females will give birth to young after a 112 to 118-day gestation period.

Species name:	*Potus flavus*
Features:	Woolly brown fur; long prehensile tail
Habitat:	Tropical forests and woodlands
Distribution:	Southern Mexico to southern Brazil
Length:	Up to 76cm (30in) without tail
Weight:	Up to 4.5kg (10lb)
Breeding:	One or two young after a 112- to 118-day gestation

CARNIVORA: PROCYONIDAE

Raccoon

The raccoon is one of nature's true survivors and opportunists, being found in almost every conceivable terrain from southern Canada to Central America. It grows up to 63cm (26in) long, with a coating of shaggy grey-black fur and deep black eye patches. The feet are black, and the tail is ringed with dark fur. Raccoons are known for their adaptability and intelligence, and they enjoy living around human habitations to access refuse food and also to make dens under buildings or any other available structure. They are omnivorous creatures, eating fruit, nuts, corn, frogs, fish, birds, eggs, insects and rodents, as well as processed human foods. The forepaws are especially nimble, and are useful for handling food – the raccoon can also open doors and unlock latches to access buildings. Raccoons are active throughout the day and night, but have superb night vision and acute hearing. They are usually solitary creatures and establish home ranges of 1–10km (0.6–6 miles) in diameter, depending upon the terrain.

Species name:	*Procyon lotor*
Features:	Grey-black fur; black eye patches; long bushy tail
Habitat:	Woodlands and forests, arid regions; marshlands and swamps, urban
Distribution:	Southern Canada to Central America
Length:	Up to 65cm (26 in) without tail
Weight:	Up to 8kg (18lb)
Breeding:	Three to seven young after a 63- to 65-day gestation

CARNIVORA: URSIDAE

Giant Panda

A dwindling population of giant pandas survive in a small pocket of southwestern China, their existence threatened by destruction of the bamboo forests. Bamboo constitutes around 99 per cent of the panda diet, with an adult bear consuming an average of 12–15kg (26–33lb) of bamboo plant material every day. It can supplement its food with carrion, birds' eggs and insects when necessary. The giant panda appearance is familiar, particularly the white head and neck featuring prominent black ears and black eye patches. Giant pandas live alone except in the case of a mother with cubs. An average home range for a male will be 8.5 sq. km (3.3 sq. miles), the female territories being roughly half this extent. Territories are marked out through urine, claw markings and anal-gland scents. Communication between pandas is accomplished by a range of 11 different vocalizations. Infant mortality is high, even among cubs born in captivity, and there is a question mark over whether the tiny giant panda population has the genetic diversity to survive.

Species name:	*Ailuropoda melanoleuca*
Features:	Black body fur, white rump; white face, black eye patches and ears
Habitat:	Bamboo and coniferous forests at 2590–3500m (8500–11,500ft) altitude
Distribution:	Southwestern China
Length:	Up to 1.9m (6ft 9in) without tail
Weight:	Up to 125kg (280lb)
Breeding:	One or two cubs after a 125- to 150-day gestation

CARNIVORA: URSIDAE

Sun Bear

The sun bear is one of the more unusual members of the *Ursidae* family. It inhabits the tropical rainforests of Southeast Asia, and has some excellent adaptations to the environment. Its claws are long and curved, while the feet pads are hairless, and this combination makes the feet superb for climbing and also for digging out insects and honey. Its tongue is also unusually long – it can reach up to 25cm (10in) from the mouth – and is used for drawing out insects from holes and honey from bees' and wasps' nests. In overall dimensions, the sun bear is the smallest of all bears, growing to a maximum of only 1.4m (4ft 6in) long in the body. It is nocturnal in habits, feeding at night and sleeping during the day (it often sleeps up in the trees). Sun bears are omnivorous; as well as insects, termites and honey, they enjoy small mammals and birds, and many agricultural crops (particularly oil palms). Little is known about the sun bear's social organization, and it is critically endangered through habitat loss and persecution.

Species name:	*Helarctos malayanus*
Features:	Black body, white to red patch on the chest; grey to orange muzzle
Habitat:	Tropical rainforests
Distribution:	Southeast Asia
Length:	Up to 1.4m (4ft 6in) without tail
Weight:	Up to 65kg (145lb)
Breeding:	One or two cubs after a 95-day gestation

CARNIVORA: URSIDAE

Sloth Bear

The sloth bear is an unusual creature with a very long, shaggy black to red-brown coat, a creamy muzzle and a distinctive white chest mark forming a U, Y, O or heart shape. It is concentrated in South Asia, particularly India and Sri Lanka, and its diet is predominantly ants, termites and fruit. It has an unusual method of raiding ant and termite nests. It breaks open the nests with its powerful front claws, then sticks in its snout and literally sucks the creatures into its mouth (it has no inner upper incisors to restrict the flow of insects into its mouth). The sucking noises are audible up to 100m (330ft) away. Available information suggests that sloth bears live solitary lives (except for during the mating season and females with cubs), but form temporary feeding groups of up to seven bears. The sloth bear's habitat is predominantly forest and grassland, although it can also survive in arid regions.

Species name:	*Melursus ursinus*
Features:	Brown to black coat; white chest marking
Habitat:	Forests, woodlands and arid scrublands
Distribution:	South Asia
Length:	Up to 1.8m (6ft) without tail
Weight:	Up to 190kg (420lb)
Breeding:	One or two cubs after a six- to seven month gestation

CARNIVORA: URSIDAE

Spectacled Bear

The spectacled bear is located in western South America, and is South America's only bear species. It takes its name from the cream-coloured markings that extend around the bear's eyes either partially or completely. The rest of the fur is black to brown, and the spectacled bear grows to around 2m (6ft 6in) in length minus the tail. Only around 2000–2400 spectacled bears remain in the wild, the declining numbers precipitated by habitat destruction and also by persecution owing to their enjoyment of agricultural crops. Spectacled bears live in a wide range of habitats, ranging from grasslands through deserts to forest. Its clash with humans, however, has pushed them mainly into montane forests above 1000m (3300ft). As well as crops, spectacled bears eat a broad range of plant foods, mashing up tough bulbs, roots and plant fibre with exceptionally powerful jaws, as well as rabbits, mice, insects, birds (and their eggs) and carrion. Females become sexually mature at the age of between four and seven years old, and produce one to three cubs.

Species name:	*Tremarctos ornatus*
Features:	Black to red-brown body; 'spectacle' markings around the eyes
Habitat:	Montane forests
Distribution:	Western South America
Length:	Up to 2m (6ft 6in) without tail
Weight:	Up to 175kg (390lb)
Breeding:	One to three cubs after a 250-day gestation

CARNIVORA: URSIDAE

American Black Bear

The American black bear inhabits woodlands, forests and montane environments across North America and down into Mexico. The name is something of a misnomer, as the uniform coat colour varies from jet black to grey-blue, the darker animals tending to live in the east. The black bear has adapted to a variety of climates because of its territorial range, from the Arctic environment of Alaska down to the arid regions of Mexico. They live alone with the exception of females with cubs, although groups of bears may feed in each other's presence if there is an abundant food source. Male home ranges can be in excess of 40 sq. km (15 sq. miles). Female home ranges are smaller, and on occasions a young female may have a sub-territory within that of its mother. The bears also hibernate throughout the winter. Black bears are omnivorous, with the vast majority of their food derived from plants. They can be determined predators, however, and will eat fish and even young deer and moose.

Species name:	*Ursus americanus*
Features:	Uniform black to blue-grey body
Habitat:	Woodlands, forests and montane environments
Distribution:	Throughout North America to Mexico
Length:	Up to 1.9m (6ft 3in) without tail
Weight:	Up to 300kg (660lb)
Breeding:	One to five cubs after a 220-day pregnancy

CARNIVORA: URSIDAE

Grizzly Bear

The grizzly bear – *Ursus arctos horribilis* – is actually a subspecies of the brown bear – *Ursus arctos*. The main distinction of the grizzly is its 'grizzled' fur texture, caused by the colour lightening at the tips of the hair. Grizzlies also tend to inhabit remote mountain and forest environments, and stay away from the coastal regions inhabited by many other brown bear subspecies. Brown/grizzly bears are immensely powerful creatures, growing up to 3m (9ft 9in) in body length with a weight of 1000kg (2200lb) – males are twice the weight of females, but of roughly the same dimensions. They have well-developed shoulder muscles (these form a pronounced hump), and long claws for digging up roots and bulbs, and for excavating dens in earth or snow. Brown bears eat intensively during the summer months in preparation for the winter sleep (they do not hibernate in the proper sense), and their carnivorous diet includes salmon, carrion and small deer.

Species name:	*Ursus arctos horribilis*
Features:	Uniform 'grizzled' brown coat
Habitat:	Woodlands and forests, and montane environments
Distribution:	Northwest North America
Length:	Up to 3m (9ft 9in) without tail
Weight:	Up to 1000kg (2200lb)
Breeding:	Usually two cubs after a 180- to 250-day gestation

CARNIVORA: URSIDAE

Polar Bear

Polar bears survive in the hostile Arctic and North Canadian environment through some amazing adaptations. The translucent guard hairs on the coat actually transfer heat from the sun down to the skin, while the underfur and blubber provide superb insulation (the blubber layer can measure 11cm/4.3in thick). The insulation, as well as the hollow (and therefore buoyant) guard hairs, allows the bear to swim in the Arctic waters, which it does fluidly, using its large paws for propulsion, taking it to speeds of 10km/h (6mph) and distances of up to 96km (60 miles). It can also stay underwater for up to two minutes. Appropriately for the climate, polar bears have an almost totally carnivorous diet – they are the world's largest land predator. The main diet is seals, which they either stalk and then charge, or drag out of the water when the seal takes a breath at a breathing hole. Other foodstuffs include lemmings, lichen and carrion. Polar bears are generally solitary, the male and females coming together to mate in April and May.

Species name:	*Ursus maritimus*
Features:	Creamy coat reflecting surrounding colours
Habitat:	Inland, coastal and aquatic environments
Distribution:	Northern Canada and Arctic
Length:	Up to 3.4m (11ft) without tail
Weight:	Up to 680kg (1500lb)
Breeding:	Usually two cubs after a 190- to 260-day pregnancy (includes period of delayed implantation)

CARNIVORA: URSIDAE

Asiatic Black Bear

The Asiatic black bear has roughly the same dimensions as the American black bear, although it tends to be lighter in weight. Its coat ranges from brown to black, and it features a prominent white to yellow patch (sometimes V-shaped) on its chest. The habitat of the Asiatic black bear is woodlands and forests, particularly montane forests (they can range up to 3000m/9900ft altitude). Diet is generally grasses, leaves, nuts, fruit, bamboo shoots, insects, small mammals and carrion. Its plant-food gathering is assisted by excellent climbing abilities, and it is also known for its bipedal confidence, walking on its rear legs easily. Like many omnivorous bears, the Asiatic black bear is known to enjoy agricultural crops. This, along with the desire for bear body parts in traditional Chinese medicine, has led to widespread bear hunting, and the Asiatic black bear is currently classified as 'vulnerable' by the IUCN. Socially the bears are nocturnal and mainly solitary, establishing home ranges of up to 20 sq. km (8 sq. miles).

Species name:	*Ursus thibetanus*
Features:	Brown to black coat; white to yellow chest patch
Habitat:	Forests and woodlands
Distribution:	South and East Asia
Length:	Up to 1.9m (6ft 3in) without tail
Weight:	Up to 200kg (440lb)
Breeding:	One to three cubs (gestation unknown)

CARNIVORA: VIVERRIDAE

Binturong

The binturong lives in the tropical forests of South and Southeast Asia. Its body length reaches up to 96cm (38in) and a subspecies, *Arctictis binturong penicillatus*, can have a body weight of up to 14kg (31lb). It is covered with a thick, shaggy fur, which is dark brown to black in colour. Its ears feature long tufts, and it has a long prehensile tail. Although the binturong is a member of the *Carnivora* order, fruit is the chief element of its diet, but it justifies its classification by also eating carrion, insects, birds, eggs, fish and rodents. Binturongs are true tree-dwelling animals, and they are seen to move very slowly along branches. Socially the animals are matriarchal, and live as parent/infant groups, although they are frequently solitary. They also territorial, and mark out arboreal home ranges using urine and gland scents. They are common creatures, but their use by humans for food and in traditional medicine means this may not always the case.

Species name:	*Arctictis binturong*
Features:	Brown to black shaggy coat; prehensile tail; tufted ears
Habitat:	Tropical forests
Distribution:	South and Southeast Asia
Length:	Up to 96cm (38in) without tail
Weight:	Up to 14kg (31lb)
Breeding:	One to three young after a 92-day gestation

CARNIVORA: VIVERRIDAE

African Civet

Civets have an appearance somewhere between that of a cat and a hyena, and the African civet is a classic representative of the group. The overall body is catlike in posture, with a slim head held low and powerful thighs and hind legs for stalking and pouncing on prey. The base colour of the coat is a creamy silver, which is covered by brown to black stripes and spots. The legs are black, and there are also dark black face markings across the eyes. Civets are secretive creatures, but are mainly nocturnal and solitary, as far as we know. They live in southern and Central Africa, and their habitat is forests and grasslands – they enjoy plenty of vegetation camouflage and like to be near sources of water. Newborn civets are relatively independent compared to other carnivores, and they can crawl at birth. After only two months they will be hunting for themselves.

Species name:	*Civettictis civeta*
Features:	Silver-cream coat covered with brown markings; black legs; black face patches
Habitat:	Forests and grasslands
Distribution:	Southern and Central Africa
Length:	Up to 91cm (36in) without tail
Weight:	Up to 4.5kg (10lb)
Breeding:	One to four young after a 60- to 72-day gestation

CARNIVORA: VIVERRIDAE

Common Genet

The common genet, also known as the small-spotted genet, inhabits a broad territorial range reaching from northwestern Europe down to the southernmost reaches of Africa. Generally it has a yellow-grey fur with dark spots over the body and dark rings around the tail, although the density of the colours varies according to the region and climate (the dark colours are seen more in the temperate zones). A short mane of hair runs along the neck and back, and this can stand up when the cat is alarmed or frightened. The common genet has a typical *Viverridae* omnivorous diet ranging from fruit down to carrion. Living prey is stalked in the manner of a cat, the genet going close and flat to the ground to close the distance from the prey, ready for a pounce and kill. Genets are solitary creatures socially, apart from during breeding, which can occur all year round (it is usually timed to correspond with times of plentiful food or during the wet season).

Species name:	*Genetta genetta*
Features:	Yellow-grey fur covered with dark spots; ringed tail; mane along neck and back
Habitat:	Woodlands, forests and grasslands
Distribution:	From northern Europe to southern Africa
Length:	Up to 55cm (22in) without tail
Weight:	Up to 2.5kg (5½lb)
Breeding:	Two or three kittens after a 70-day gestation

CARNIVORA: VIVERRIDAE

Large-spotted genet

The large-spotted genet's difference from the small-spotted genet (common genet) is implied in its name. Base fur colour ranges from grey through to a light brown, and the coat is decorated with large rust-brown spots (it is also called the rusty-spotted genet). The tail is a pale creamy colour and is ringed with brown markings; the tip is black. White patches appear on the face running beneath the eyes. In terms of dimensions, behaviour and reproduction, there is little to separate this genet from the common genet. Its distribution, however, is much narrower, being found purely in southern and eastern Africa. There it gravitates towards river areas, using the thick riverbank foliage to move while remaining concealed. The genet tends to rest up during the day, sleeping in a tree or in an abandoned aardvark burrow. Genets have a mixed relationship with people – they are appreciated for their rodent control, but are also killed for the devastation they can wreak on poultry.

Species name:	*Genetta tigrina*
Features:	Grey to light brown fur covered with large, rust-coloured spots; ringed tail; mane along neck and back; white patches beneath eyes
Habitat:	Woodlands, forests, grasslands and riverbanks
Distribution:	Southern and eastern Africa
Length:	Up to 55cm (22in) without tail
Weight:	Up to 2.5kg (5½lb)
Breeding:	Two or three kittens after a 70-day gestation

CARNIVORA: VIVERRIDAE

Dwarf Mongoose

Dwarf mongooses are highly sociable animals, living together in packs numbering as many as 20, but more commonly 12–15. The pack will establish a territory of around 30 hectares (75 acres), and this will contain numerous termite mounds. Termites are the principal part of the dwarf mongoose diet, but they will take many other types of insect, as well as mice (and other rodents), lizards, young birds and fruit. Within the range, the pack will keep moving its den location, rotating between the termite mounds to keep the food supply strong. Dwarf mongooses are diurnal in behaviour, and are most active in the cooler morning hours, when the animals will emerge to play, groom themselves or feed. Within the pack will be a dominant breeding pair, and a female is capable of producing up to 12 offspring each year. The dwarf mongoose has a uniform red-black/brown speckled fur and, as the name implies, its dimensions are small, reaching a maximum body length of around 28cm (11in).

Species name:	*Helogale parvula*
Features:	Red-black/brown speckled fur; long claws on front feet
Habitat:	Grasslands, forests and semi-arid areas
Distribution:	Southern and eastern Africa
Length:	Up to 28cm (11in) without tail
Weight:	Up to 350g (13oz)
Breeding:	Two to four young after a 50- to 54-day gestation

CARNIVORA: VIVERRIDAE

Mongoose

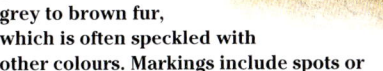

The mongoose comes in more than 40 different species, with those of the *Herpestes* genus being most widespread. Despite some distinct visual differences in certain regions, the animals of the mongoose family have common characteristics. They generally have short grey to brown fur, which is often speckled with other colours. Markings include spots or stripes, depending upon the species. Mongooses are most famous for their powers of predations, particularly in attacking and killing lethally poisonous snakes. Tackling such animals is high risk for the mongoose – they do not have immunity to the venom – but their lightning-fast reflexes enable them to avoid the snake strike, then bite through the snake's skull to kill it. The mongoose diet consists of small mammals, reptiles, birds and their eggs, and some fruit. Social life varies enormously, with some mongooses living solitary lives, while others gather in large packs (*see* dwarf mongoose, opposite). Mongooses are mostly diurnal animals, and like terrain with plenty of foliage cover and opportunistic dens.

Species name:	*Herpestes edwarsi* (data for Indian Grey mongoose)
Features:	Grizzled grey to brown fur; reddish colours on feet and head; tail as long as body
Habitat:	Grasslands, forests and semi-arid areas
Distribution:	South Asia
Length:	Up to 45cm (18in) without tail
Weight:	Up to 1.7kg (4lb)
Breeding:	Two to five young after a 60- to 65-day gestation

CARNIVORA: VIVERRIDAE

Banded Mongoose

The banded mongoose is so-called from the dark brown bands that cross over the back of the animal's grizzled brown-grey-white coat. Like the dwarf mongoose, banded mongooses are pack animals, forming groups of between 4 and 40 members, with a dominant male at the top of the hierarchy. The pack will embark on group food-gathering trips that can extend over 8km (5 miles) in the day. Territory is marked out using emissions from anal glands and cheek glands, and the mongooses are also known to cover each other with scent when reunited after a predator attack or similar scare. Their strong pack instinct has even led to the animals attempting to rescue other mongooses caught by larger and more powerful predators. The banded mongoose nests and sleeps in burrows, which it either digs itself with its long, non-retractable claws or adapts from termite mounds or rodent/mammal holes. Pregnant female banded mongooses have a compressed gestation period of two months, after which a litter of two to six blind and helpless young are born.

Species name:	*Mungos mungo*
Features:	Grizzled brown-grey-white fur; dark bands across the back
Habitat:	Grasslands, woodlands and semi-arid areas
Distribution:	Sub-Saharan Africa
Length:	Up to 45cm (18in) without tail
Weight:	Up to 2.5kg (5lb 8oz)
Breeding:	Two to six young after a 60- to 65-day gestation

CETACEA: BALAENIDAE

Northern Right Whale

The northern right whale is an endangered species that inhabits many of the waters of the northern hemisphere (*Eubalaena glacialias*, the southern right whale, lives in the southern hemisphere). Its vulnerability extends from spending much of its time at the surface, where it is exposed to hunting and boat strikes. The right whales are baleen whales, meaning that their mouths contain two rows of baleen plates, which filter out plankton and small crustaceans from the water as food. They are huge creatures, and the northern right whale can grow up to 17m (56ft) in length, with a weight of 80 tons (88 tonnes). The skin is grey-black in colour, with occasional white patches and large growths known as 'callosities' around the head area. There is no dorsal fin, and two blowholes are positioned on top of the head. Socially, the right whales may form into feeding groups, but the most typical social unit is a mother and offspring pair.

Species name:	*Eubalaena glacialis*
Features:	Grey-black skin; white patches; callosities around head
Habitat:	Temperate and sub-polar waters
Distribution:	Northern hemisphere
Length:	Up to 17m (56ft)
Weight:	Up to 80 tons (88 tonnes)
Breeding:	One calf born after a 12-month gestation

CETACEA: BALAENOPTERIDAE

Blue Whale

The blue whale is the world's largest animal. It can grow up to 30m (98ft) in length and weigh up to 160 tons (176 tonnes). Females are generally larger than the males. Its huge size is balanced by a relatively slender overall build (although it tends to be fatter in the summer months). The skin is blue-grey and is covered by white-grey spots. Up to 70 grooves run along the skin on the underside, from the mouth to several metres behind the flippers; the flippers themselves are around 2.4m (8ft) long, and the blue whale also has a very small dorsal fin. Blue whales have been referred to as 'gulpers' in feeding technique – the whale enters a shoal of prey, takes a huge mouthful of water and food (the throat at this point expands by up to 400 per cent), then pushes the water through the baleen plates to sieve out the prey. Blue whales can make powerful vocalizations of up to 180 decibels (a jet engine is typically 140 decibels), and they can dive to depths of 200m (655ft).

Species name:	*Balaenoptera musculus*
Features:	Blue-grey skin with pale grey spotting; 55–70 grooves along underside of neck and body
Habitat:	Deep offshore waters
Distribution:	Worldwide (migrate from tropical to temperate waters in summer)
Length:	Up to 30m (98ft)
Weight:	Up to 160 tons (176 tonnes)
Breeding:	One calf born after a 12-month gestation

CETACEA: BALAENOPTERIDAE

Humpback Whale

The humpback whale is as much identified through its behaviour as its appearance. Its most spectacular trait is 'breaching' – the whale dives down, then swims upwards at great speed to exit the water entirely before twisting around and smashing back onto the water surface. There is no comprehensive explanation to this action, but theories include play behaviour, sound-wave generation and the loosening of skin parasites. Humpback whales are also known for the song of the male, who emits a rich, haunting song for up to 30 minutes, possibly as a mating call or as a sonar location device. Other interesting behaviours (not necessarily exclusive to the humpback) include skyhopping – poking the head out of the water to look around – and lobtailing – slapping the surface of the water with the tail. In appearance, humpback whales are up to 14m (46ft) long with a dark blue to pale grey skin colour, a grooved throat and extremely long flippers – these can be one-third of the whale's body length.

Species name:	*Megaptera novaeangliae*
Features:	Dark-blue to grey skin; throat grooves; tubercles around front jaws; long flippers
Habitat:	Coastal and offshore waters
Distribution:	Worldwide
Length:	Up to 14m (46ft)
Weight:	Up to 30 tons (33 tonnes)
Breeding:	One calf born after a 11- to 12-month gestation

CETACEA: DELPHINDAE

Common Dolphin

Common dolphins, as with all other species of dolphin, belong to the *Odentoceti* suborder of the *Cetacea* family – they are toothed whales. They have an almost worldwide distribution through temperate and tropical waters. While the dolphin's back is grey to black, the flanks have a colourful yellow and pale-grey 'hourglass' pattern, the yellow section extending from the face to just beneath the dorsal fin. Recently the common dolphin has also been divided into two species: the short-beaked common dolphin (*Delphinus delphis*) and the long-beaked dolphin (*Delphinus capensis*). As with most other dolphins, common dolphins are social animals, forming schools of thousands of animals that cooperate in feeding activities (schools of more than 10,000 dolphins have been estimated). The diet is formed of schooling fish and squid, and the dolphins will often herd the shoals of fish in small areas to intensify the feeding. Common dolphins are graceful and powerful aquatic animals, and can dive to 200m (656ft).

Species name:	*Delphinus delphis; Delphinus capensis*
Features:	Grey to black skin on back; yellow and grey hourglass pattern on flanks
Habitat:	Temperate and tropical waters
Distribution:	Worldwide (within habitat types)
Length:	Up to 2.6m (8ft 6in)
Weight:	Up to 80kg (175lb)
Breeding:	One calf born after a 10- to 11-month gestation

CETACEA: DELPHINDAE

Pilot Whale

Pilot whales are found in two species: *Globicephala melas* (long-finned pilot whale) and *Globicephala macrorhynchus* (short-finned pilot whale) – the names explain the distinction between the two. The shared appearance is a grey–black skin colour, a prominent, rounded dorsal fin and – most distinctive – a swollen head with a forehead that can project up to 10cm (4in) in front of the lower jaw. The flippers are sickle-shaped. Pilot whales are highly intelligent creatures, and the US Navy has experimented with training the animals for operational tasks. Socially they form schools typically numbering up to 90 creatures, but on occasion the schools can be in the hundreds. Vocal clicks and whistles enable the whales to recognize individuals and to communicate, and groups of pilot whales are known to 'herd' shoals of fish when feeding. The mother–offspring bond is very tight, with the nursing phase extending up to and over 22 months. Male pilot whales fight each other for breeding females with the bites inflicting substantial scars on the skin.

Species name:	*Globicephala melas; Globicephala macrorhynchus*
Features:	Grey to black skin; bulbous forehead
Habitat:	Temperate and tropical waters
Distribution:	Worldwide (within habitat types)
Length:	Up to 7m (23ft)
Weight:	Up to 1.8 tons (2 tonnes)
Breeding:	One calf born after a 12- to 15-month gestation

CETACEA: DELPHINDAE

Orca

The orca, or killer whale, has an instantly recognizable black-and-white colour scheme, the body being black except for a white underside, white eye patches and a grey saddle patch behind the dorsal fin. Tail, flippers and dorsal fins are very large, giving the orca the propulsion and manoeuvrability it needs for hunting the fish, smaller whales, porpoises, sharks, seals, turtles, penguins and even birds that form part of its varied diet. Working in groups (called 'pods') of between 20 and 150 animals, orcas have a wide range of hunting methods. These include drowning smaller whales by forcing them underwater to prevent breathing and beaching themselves on land to grab seals and penguins. Despite their undoubted ferocity, orcas are highly social animals. Mother and calf bonds are intense and can be lifelong (pods are matriarchal by order), and when the pods are on the move the mothers and young are placed at the centre for protection. Injured orcas are often fed by other members of the orca group.

Species name:	*Orcinus orca*
Features:	Black skin with white underside, white eye patches and grey saddle patch
Habitat:	All ocean waters
Distribution:	Worldwide
Length:	Up to 9m (30ft)
Weight:	Up to 10 tons (11 tonnes)
Breeding:	One calf born after a 13- to 17-month gestation

CETACEA: DELPHINDAE

Spotted Dolphin

There are several species of spotted dolphin, two of the most common being dealt with here. The pantropical spotted dolphin (*Stenella attenuata*) is the larger of the pair, growing up to 2.6m (8ft 6in). It is found worldwide in tropical and temperate waters, within which it is common. From the top of the head to just behind the dorsal fin the colouration is grey, with a lighter grey section running along the flanks beneath and white underparts. Contrasting spots cover most of the body. The Atlantic spotted dolphin (*Stenella frontalis*) inhabits only Atlantic waters. It has a similar colour scheme to its pantropical cousin, but its body is shorter and less slender. In all of the spotted dolphin types, the number and colour of the spots change with habitat and age. Generally, older dolphins or those living near the coast have more and darker spots than younger dolphins or those that inhabit deep oceanic waters. The data below apply to *Stenella attenuata*.

Species name:	*Stenella attenuata*
Features:	Body divided longitudinally into dark grey, light grey and white sections; contrasting spots covering the body
Habitat:	Tropical and temperate waters
Distribution:	Worldwide
Length:	Up to 2.6m (8ft 6in)
Weight:	Up to 120kg (260lb)
Breeding:	One calf born after a 11½-month gestation

CETACEA: DELPHINDAE

Bottlenose Dolphin

The bottlenose dolphin is a familiar resident of many marine zoos, it being a social and highly intelligent creature. Its colouration shifts from dark grey or black on its back to white on the underparts, and its name relates to its relatively short, rounded beak. One of its distinctive physical features is its vertebrae – five of its seven neck vertebrae are not fused together, which means that it has greater neck flexibility than many other dolphin species. Bottlenose dolphins have a typical dolphin diet consisting of fish, squid, crustaceans and other marine wildlife. Hunting of large shoals of fish often takes place within a group context. Dolphin groups range in number from around a dozen to several hundred, the larger groups being found in the open ocean (the bottlenose dolphins in these environments, particularly in cold waters, are also much bigger than those specimens found in warmer coastal regions). Bottlenose dolphins are highly communicative, using noises ranging from vocal clicks to water slaps within the group.

Species name:	*Tursiops truncatus*
Features:	Dark grey to white colouration; long, curved dorsal fin; short beak
Habitat:	All waters except polar
Distribution:	Worldwide
Length:	Up to 4m (13ft)
Weight:	Up to 500kg (1100lb)
Breeding:	One calf born after a 12-month gestation

CETACEA: ESCHRICHTIDAE

Grey Whale

Only an international protection agreement in 1937 saved the grey whale from complete extinction, but even today the animal remains critically endangered. Current global population estimates are around 17,000 animals. Its skin is a mottled stony colour, with 6–12 bumps (of diminishing size) along the dorsal section instead of an actual dorsal fin. The head is slender and tapering, and this and the rest of the body are frequently clustered with barnacles and other sea parasites. Grey whales are baleen whales, and one of their feeding techniques is to scoop mud from the seabed (the mud is rich in foods such as tubeworms). The grey whale is known for making the longest migration of any animal on the planet. It heads up to the Arctic waters of the Bering and Chukchi seas during the summer months to feed, while going south to the waters of the East Pacific in the winter to mate and calve. The total distance of the round trip is up to 22,500km (14,000 miles).

Species name:	*Eschrictius robustus*
Features:	Grey body; series of dorsal bumps; notched flukes
Habitat:	Open ocean; coastal waters
Distribution:	Pacific
Length:	Up to 15m (49ft)
Weight:	Up to 35 tons (38.5 tonnes)
Breeding:	One calf born after a 12- to 13-month gestation

CETACEA: MONODONTIDAE

Beluga

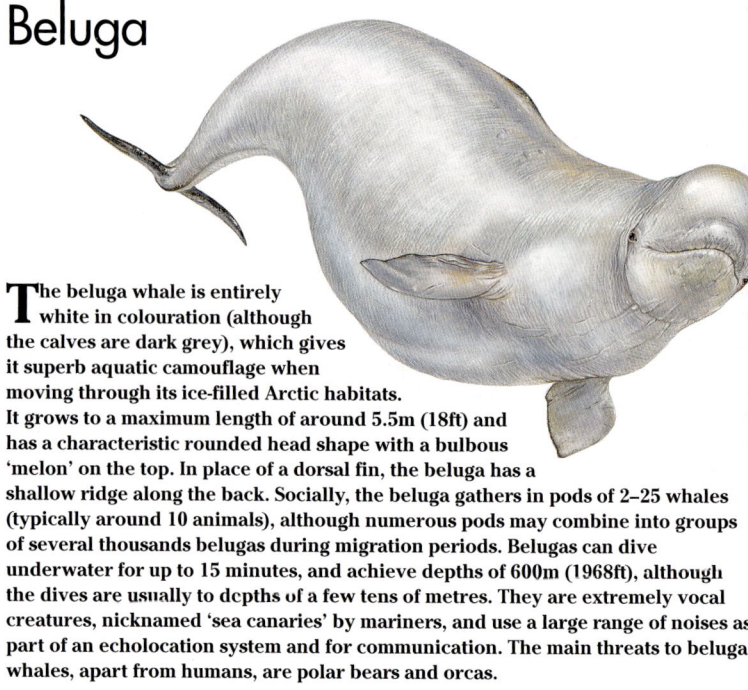

The beluga whale is entirely white in colouration (although the calves are dark grey), which gives it superb aquatic camouflage when moving through its ice-filled Arctic habitats. It grows to a maximum length of around 5.5m (18ft) and has a characteristic rounded head shape with a bulbous 'melon' on the top. In place of a dorsal fin, the beluga has a shallow ridge along the back. Socially, the beluga gathers in pods of 2–25 whales (typically around 10 animals), although numerous pods may combine into groups of several thousands belugas during migration periods. Belugas can dive underwater for up to 15 minutes, and achieve depths of 600m (1968ft), although the dives are usually to depths of a few tens of metres. They are extremely vocal creatures, nicknamed 'sea canaries' by mariners, and use a large range of noises as part of an echolocation system and for communication. The main threats to beluga whales, apart from humans, are polar bears and orcas.

Species name:	*Delphinapterus leucas*
Features:	Adults have white/blue-white body; bulbous forehead; no dorsal fin
Habitat:	Arctic waters
Distribution:	Arctic Ocean
Length:	Up to 5.5m (18ft)
Weight:	up to 1.5 tons (1.7 tonnes)
Breeding:	One calf born after a 14- to 15-month gestation

CETACEA: PHYSETERIDAE

Sperm Whale

The sperm whale is a massive animal (the world's largest toothed whale), growing up to 20m (65ft) in length and weighing up to 57 tons (62.8 tonnes). It has a huge, rectangular-shaped head, brown-grey, wrinkly skin with pale underparts and jawline, and a series of ridges between the dorsal fin and the enormous tail. The head contains the spermaceti organ, which produces and stores the waxy spermaceti oil. The oil, which has unfortunately attracted whalers to the sperm whale, changes in consistency with depth and temperature, and may provide buoyancy to the whale when diving. Sperm whales can dive very deep indeed, with evidence of dives below 3000m (9842ft). On average, however, sperm whales dive for between 20 and 50 minutes and reach depths of 600m (1968ft). No light penetrates to such depths, so the sperm whale relies on echolocation to finds its way and detect prey. The main diet of sperm whales is squid, octopus and fish, and around 900kg (1980lb) of food is consumed every day.

Species name:	*Physeter catodon*
Features:	Square head shape; brown-grey wrinkled body; white underparts
Habitat:	Deep oceanic waters
Distribution:	Worldwide
Length:	Up to 20m (65ft)
Weight:	Up to 57 tons (62.8 tonnes)
Breeding:	One calf born after a 14- to 16-month gestation

CHIROPTERA: MEGADERMATIDAE

Ghost Bat

The ghost bat, also known as the Australian false vampire bat, inhabits the full range of environments on the northern coastline of Australia, from desert outback to tropical rainforests. Its name derives from its 'ghostly' colouration – light blue-grey to white bodies (white underparts) – while its alternative name acknowledges the popular myth that the bat drinks blood as part of its diet. Ghost bats are fairly small creatures, averaging around 10cm (4in), and they have proportionately large ears and a noseleaf. They are also Australia's only carnivorous bat, dieting on insects, birds, frogs, small mammals and other bats. As with most bats, it detects prey in the dark via echolocation. Through much of the year, the ghost bat roosts on its own or in a small group, but during the breeding species larger sex-separated colonies are formed. Typical roosting sites are caves, mines and quarries. Habitat disturbance by humans is a major threat to the ghost bat, and it is classified as a 'vulnerable' species by the IUCN.

Species name:	*Macroderma gigas*
Features:	Blue-grey upper body; white underparts; long ears; noseleaf
Habitat:	Tropical terrains ranging from arid zones to rainforest
Distribution:	Northern Australia
Length:	Up to 12cm (5in)
Weight:	Up to 150g (5oz)
Breeding:	One infant born after a 3-month gestation

CHIROPTERA: MOLOSSIDAE

Mexican Free-tailed Bat

Free-tailed bats are so called because the tail is not incorporated into the tail membranes, and so hangs free. They are to be found in a wide range of habitat types from the southern United States down to southern Brazil and Argentina, and they make long migratory flights to breeding grounds, especially between Mexico and Texas. Their colonies are huge – often numbering in the millions (one colony near San Antonio, Texas, contains around 20 million creatures) – and they roost during the daytime hours before flying out at night in streams of creatures to feed. Thousands of pups are left behind in the roosts when the adults feed, but the mothers have no difficulty in locating their offspring by each pup's distinctive cries. The vast feeding parties have a significant pest control impact on local areas, particularly in the reduction of moths and flies. Mexican free-tailed bats grow to around 10cm (4in) long and have red-brown to grey fur and wide-set ears (an advantage for echolocation).

Species name:	*Tadarida brasiliensis*
Features:	Red-brown to grey fur; wide-set ears; naked 'free' tail
Habitat:	Caves, under bridges and in domestic dwellings
Distribution:	Southern USA down to Argentina and Chile
Length:	Up to 10cm (4in) without tail
Weight:	Up to 30g (1oz)
Breeding:	One pup born after a 11- to 12-week gestation

CHIROPTERA: NOCTILIONIDAE

Greater Bulldog Bat

The greater bulldog bat – so called because its prominent cheek pouches and folds of skin around the mouth evoke the bulldog's expression – feeds mainly upon fish (it is also called the fish-easting bat). Using echolocation, the bat detects shoals of fish beneath the river and coastal waters of Central and South America, then hones in on the ripples made by the fishes' swimming. Diving down, the bulldog bat tucks in its interfemoral membrane and sweeps its feet through the shoal to a depth of around 2.5cm (1in), grabbing a fish as it comes into contact with it. A large, broad wingspan enables the bulldog bat to glide over the water when hunting, and not alarm the fish by flapping. Much of the bat's body is hairless, the exception being the head, shoulders and back, which have a covering of short orange-brown fur with a pale stripe along the middle of the back. Bulldog bats roost in trees and in rocky crevices, and hunt mainly during the dusk and night-time hours.

Species name:	*Noctilio leporensis*
Features:	Nose tip projecting over nostrils; fleshy folds of skin around lips; orange-brown fur on head, shoulders and back
Habitat:	Tropical forests, caves, trees
Distribution:	Central and South America
Length:	Up to 8cm (3in) without tail
Weight:	Up to 35g (1.2oz)
Breeding:	One pup born after a 16-week gestation

CHIROPTERA: PHYLLOSTOMIDAE

Vampire Bat

The small size of the vampire bat – its maximum length is around 9.5cm (3.7in) and its wingspan averages 20cm (8in) – does not reflect the magnitude of its social reputation. Its fur is brown-grey and it has a very squat, flattened nose. The forearms and legs are very well developed, and it is quite confident moving at speed across land. To feed, it creeps up on a warm-blooded animal and uses its literally razor-sharp teeth to cut off a portion of skin several millimetres long. It then drinks the flowing blood, typically consuming around 25ml (1fl oz) over a period of 30 minutes to 4 hours. Chemicals in the bat's saliva both prevent the blood clotting and numb the wound site to prevent the animal (or occasionally human) from waking up. The vampire bat is common throughout South American and up into Mexico, but its damage to farm livestock is leading to increased persecution and, through misidentification of species, to the killing of many other types of bat.

Species name:	*Desmodus rotundus*
Features:	Grey-brown fur with paler underparts; flattened nose; well-developed legs
Habitat:	Forests and woodlands, arid regions, grasslands and urban areas
Distribution:	Mexico down through South America
Length:	Up to 9.5cm (3.7in) without tail
Weight:	Up to 45g (1.6oz)
Breeding:	One infant born after a 210-day gestation

CHIROPTERA: PTEROPODIDAE

Indian Flying Fox

The Indian flying fox is a huge member of the bat species – its body length reaches up to 30cm (12in) and its wingspan can be 127cm (50in). It is exclusively a fruit feeder, flying up to 48km (30 miles) from its roost to find fruits such as mangoes, bananas, guavas and figs. The fruit diet means that echolocation is not as important as for insect-eating bats, and the flying fox relies heavily on sight and smell. The Indian flying fox's diet makes commercial fruit plantations popular feeding sites, and for this reason the animal is often treated as a pest and killed, although in southern parts of India many communities accord the flying fox a sacred status. When roosting, the Indian flying fox is usually found hanging upside down from a tree branch at a communal site (called a 'camp'). Newborn Indian flying foxes are well developed, with a good fur covering and open eyes. They will remain dependent upon the mother, however, for around eight months.

Species name:	*Pteropus giganteus*
Features:	Pointed muzzle; reddish-brown fur
Habitat:	Tropical forests and swamps
Distribution:	India
Length:	Up to 30cm (12in)
Weight:	Up to 1.8kg (4lb)
Breeding:	One infant born after a 140- to 150-day gestation

CHIROPTERA: RHINOLOPHIDAE

Greater Horseshoe Bat

In total, there are 62 species of horseshoe bat. Greater horseshoe bats have an overall light brown-grey fur with red overtones, and a characteristic noseleaf that has a horseshoe shape (hence its name). Noseleafs perform an important role in the bat's echolocation system, acting as an amplifier to the ultrasonic signals projected into the air. While its cousin, the lesser horseshoe bat, grows to around 4cm (1.6in), the greater horseshoe bat extends up to 7.5cm (3in), although it has a similar distribution across Europe, North Africa and western Asia. Numbers in the United Kingdom are critically low following decades of habitat loss and pesticide usage, which reduced the population by 98 per cent during the twentieth century. Greater horseshoe bats like woodlands and open fields for habitats, roosting in colonies of up to 600 animals in caves or similar concealed locations (they roost hanging upside down, with their wings wrapped around them). They emerge at dusk and sometimes dawn to feed, catching most of their prey on the wing.

Species name:	*Rhinolophus ferrumequinurr*
Features:	Light brown-grey fur; horseshoe-shaped noseleaf
Habitat:	Woodlands, forests, grasslands and arid regions; roost in caves or crevices
Distribution:	Europe, North Africa and western Asia
Length:	Up to 7.5cm (3in) without tail
Weight:	Up to 35g (1.2oz)
Breeding:	One infant born after a 75-day gestation

CHIROPTERA: VESPERTILIONIDAE

Big Brown Bat

The big brown bat has a total length of around 12cm (5in) – rather diminutive when considering its name. It has a body covered with red-brown fur, and its wings, which have a spread of around 30cm (12in), are black and leathery. While the bat inhabits forests, it is also very common around human habitations, often seen over city parks and gardens, and around farms, or roosting under bridges or in the roofs of houses. Like many bats, the big brown bat hunts on the wing using echolocation, and its typical diet is moths, dragonflies, wasps, mosquitoes, flies and many other flying insects. The bat is also very fast in flight, reaching speeds of up to 64km/h (40mph). With a body around 7.6cm (3in) long, the big brown black has a rough chestnut-brown fur and wide-set, large ears. Flying insects are an important part of its diet, but it also eats ground-dwelling insects such as ants and beetles. (Farmers particularly like the big brown bat for its aggressive consumption of cucumber beetle larvae.) Its habitats are forests and urban buildings, with trees being favourite roosting places.

Species name:	*Eptesicus fuscus*
Features:	Chestnut-brown fur; paler fur on the underparts
Habitat:	Woodlands, forests and urban areas
Distribution:	Northern Canada down to southern Mexico
Length:	Up to 7.6cm (3in)
Weight:	Up to 45g (1.6oz)
Breeding:	One or two pups born after a 75-day gestation

CHIROPTERA: VESPERTILIONIDAE

Mouse-eared Bat

Mouse-eared bats are found throughout Eurasia and extend into North Africa. They average around 7cm (3in) in length and have particularly long ears. They are also very common creatures (bats actually form a quarter of all mammal species) with a worldwide distribution, except for the polar regions and some isolated oceanic islands. The colour of the long fur varies according to species and habitat, but is generally a grey-brown or black-brown colour with paler underparts. Their wide distribution means that, mouse-eared bats are found in a huge range of habitats, from deserts to tropical rainforests. They are even found in mountains, living up to 3000m (9842ft) in altitude. Roosting sites are those typical of bats: trees, caves, houses and rock crevices. In some areas, the bats will hibernate or sleep through the cold winter months. The males and females separate during the breeding season, with the females forming huge nursing colonies for the raising of their young.

Species name:	*Myotis myotis*
Features:	Grey-brown or black-brown colour with paler underparts
Habitat:	Woodlands, forests, arid regions, tropical regions, urban
Distribution:	Worldwide except polar
Length:	Up to 7cm (3in) without tail
Weight:	Up to 45g (1.6oz)
Breeding:	One pup born after a 50- to 75-day gestation

CHIROPTERA: VESPERTILIONIDAE

Pipistrelle Bat

Pipistrelle bats are found in a large number of species in Eurasia and North Africa (mainly Morocco). The common pipistrelle (*Pipistrellus pipistrellus*) is a very small creature, the adult bat having an average body length of 3–5cm (1.2–2in). The legs are short, and the head is flat with broad ears. Fur colour is various shades of brown, while the ears and muzzle are dark-brown to black. Pipistrelles are a common sight in the dusk hours of spring, summer and autumn (they hibernate in the winter months), during which they will hunt insects – mainly moths, gnats and flies – on the wing with a characteristic jerky, highly manoeuvrable flight pattern and an altitude of around 6m (20ft). They form large colonies in buildings, trees and rock crevices, and the numbers of pipistrelle bats in hibernation colonies can number as many as 100,000 bats. Pipistrelle pups are usually born around June, the length of pregnancy controlled by a period of delayed implantation to ensure good food supplies for the new mother.

Species name:	*Pipistrellus pipistrellus*
Features:	Brown fur; flat head shape; broad ears
Habitat:	Woodlands, forests and human habitations
Distribution:	Europe and western Asia, North Africa
Length:	Up to 4.5cm (1.7in) without tail
Weight:	Up to 8g (0.3oz)
Breeding:	One pup born after a 44- to 50-day gestation (following period of delayed implantation)

DERMOPTERA: CYNOCEPHALIDAE

Colugo

Colugos – more usually known as flying lemurs (although they are not members of the lemur family) – exist as two species in the *Cynocephalidae* family, the most common being *Cynocephalus variegatus*, the Malayan flying lemur. The unique feature of the colugo is the patagium, a skin membrane which is stretched between all limbs, the neck and the tail. When extended, the patagium forms a broad wing that enables the colugo to glide across its arboreal habitats to distances of 100m (328ft) – they can cover 70m (230ft) without height loss. The Malayan flying lemur inhabits the tropical rainforests of Southeast Asia. It has a grey-brown fur featuring an irregular pattern of white markings. The head is small but features proportionately large, rounded eyes. It feeds at night on a herbivorous diet, and moves through the jungle canopy by first climbing up a tree using its claws, then gliding between trees using its patagium. Colugos tend to live alone or in single-figure groups.

Species name:	*Cynocephalus variegatus*
Features:	Grey-brown fur; white spots and patches; patagium membrane
Habitat:	Tropical forests
Distribution:	Southeast Asia
Length:	Up to 42cm (16.5in)
Weight:	Up to 2kg (4lb 6oz)
Breeding:	One infant (rarely two) born after a 60-day gestation

HYRACOIDEA: PROCAVIIDAE

Tree Hyrax

There is a total of eight hyrax species throughout Africa and the Middle East. Their appearance is similar to that of a rabbit with very short ears (they are also called the 'rock rabbit'), although genetically they have more in common with hoofed mammals and elephants. Hyraxes have compact bodies, with a thick covering of luxurious grey-brown fur with paler underparts. The legs are short but strong, and there are opposable toes on the back feet to make the hyrax a proficient climber. Tree hyraxes are somewhat larger than rock hyraxes (*see separate entry*), growing up to 73cm (29in) including the tail, and are found in eastern and southern Africa. They tend to form groups no larger than a pair, and they live and feed mainly in the trees, establishing nests in tree holes. The diet of the tree hyrax is similar to that of the rock hyrax, although it contains less foods from the floor level, mainly consisting of leaves, fruit and birds eggs.

Species name:	*Dendrohrax arboreus*
Features:	Thick grey-brown fur; paler underparts; small rounded ears
Habitat:	Woodlands, forests and shrublands, including those in mountainous terrain
Distribution:	Eastern and southern Africa
Length:	Up to 70cm (28in) without tail
Weight:	Up to 4.5kg (10lb)
Breeding:	Two or three young born after a seven-month gestation

HYRACOIDEA: PROCAVIIDAE

Rock Hyrax

There are two species of rock hyrax, and they closely resemble the appearance and habits of the tree hyrax (*see* separate entry). Rock hyraxes inhabit rocky, mountainous areas of southern and East Africa and West Asia, the opposable toes on their back feet and sticky pad secretions making them sure-footed creatures. Although hyraxes possess a good quality of fur, they have poor body temperature regulation, and so are often seen pressing against one another for warmth. Rock hyraxes are far more sociable than tree hyraxes, and form themselves into colonies of up to 50 creatures led by a dominant male. The animals communicate among themselves with a varied range of vocalizations, including an alarming array of shrieks and screams. The hyrax is a food for several different predators, including leopards, birds of prey, pythons and civets. They consequently feed cautiously, never moving far from a rocky escape hole. Hyrax young are physically independent, being able to run within an hour of birth.

Species name:	*Procavia capensis*
Features:	Thick grey-brown fur; paler underparts; small rounded ears
Habitat:	Grasslands, scrub, rocky and mountainous territories
Distribution:	Southern and East Africa and West Asia
Length:	Up to 58cm (23in) without tail
Weight:	Up to 5kg (11lb)
Breeding:	Two or three young born after a seven-month gestation

INSECTIVORA: ERINACEIDAE

European Hedgehog

European hedgehogs are common and much-loved animals throughout western Europe (they have also been introduced into New Zealand). The hedgehog name is derived from their piglike activity of rooting through undergrowth for food while snuffling noisily. Their appearance is familiar – most of the upper body and flanks are covered with closely packed spines of brown and white coloration, the spines being a form of toughened hair. An adult hedgehog can have a covering of 5000 such spines, and the animal rolls up in a ball to present the spines outwards as a form of self-defence. Hedgehogs are agile creatures that make good swimmers and climbers. They are non-territorial, and go on solitary feeding trips to track down slugs, earthworms, snails, spiders, beetles and birds' eggs. They inhabit woodlands, hedgerows, grassy fields and urban areas (they are commonly seen on garden lawns at night), and hibernate during the winter in nests of leaves and grass, waking occasionally on milder nights to feed.

Species name:	*Erinaceus europaeus*
Features:	Dense covering of brindled spines
Habitat:	Woodlands, hedgerows, grassy fields, urban areas
Distribution:	Western Europe
Length:	Up to 27cm (10.6in)
Weight:	Up to 1kg (2lb 3oz)
Breeding:	Four or five young born after a 31- to 35-day gestation

INSECTIVORA: ERINACEIDAE

Desert Hedgehog

The desert hedgehog is able to survive in the harshest of arid regions, with its range extending from Saharan North Africa into the desert lands of the Middle East. It follows the typical habits of many smaller animals in desert regions, sleeping for most of the day in a protected location (usually in rocky crevices) and hunting at night. Its diet is varied, mainly insects, lizards, frogs, snakes, scorpions, birds' eggs and some small mammals. The desert hedgehog is very resistant to both snake and scorpion venom. When attacked by predators, it adopts the standard rolled-up spines-outward posture, although it leaves itself exposed when sleeping by lying on its side. Newborn desert hedgehogs are already equipped with their spines, although these are contained under the skin during birth. Desert hedgehogs will often go into aestivation (a dormancy that reduces metabolic activity) during the summer months to survive the worst of the desert heat.

Species name:	*Paraechinus aethiopicus*
Features:	Pale brown spines with dark tips; black muzzle and face with light forehead markings
Habitat:	Deserts, scrub and grasslands
Distribution:	North Africa and Middle East
Length:	Up to 23cm (9in)
Weight:	Up to 700g (1lb 8oz)
Breeding:	Four to six young born after a 32-day gestation

INSECTIVORA: SOLENODONTIDAE

Solenodon

The *Solenodontidae* family contains two species – the Cuban solenodon (*Solenodon paradoxus*) and the Hispaniolan solenodon (*Solenodon paradoxus*). The territory of the Hispaniolan solenodon is Haiti and the Dominican Republic, while the Cuban solenodon is found in the country of its name. Both are like large shrews, with dark reddish-brown fur, long pointed snouts, well-developed limbs for digging and long tails. The hind legs, feet, nose and tail are almost bald. The Cuban animal is the larger of the two, growing up to 39cm (15in). The solenodons are now 'endangered' according to the IUCN, their numbers massively reduced by human killing (they are targeted by agricultural communities because of crop destruction) and also by predation from increased numbers of domestic cats and dogs. The solenodon is a nocturnal creature, and finds nests for itself and for family groups in hollow logs, caves, rock crevices while also excavating its own underground burrows in the soil. The data below are for *Solenodon paradoxus*.

Species name:	*Solenodon paradoxus*
Features:	Red-brown fur; long proboscis; exposed skin on nose, feet, hind legs and tail
Habitat:	Woodlands and forests
Distribution:	Oriente Province in Cuba
Length:	Up to 39cm (15in)
Weight:	Up to 1kg (2lb 3oz)
Breeding:	One or two young born after a 50-day gestation

INSECTIVORA: SORICIDAE

Northern Short-tailed Shrew

The northern (or North American) short-tailed shrew has unusually small eyes and ears – both are buried beneath the short grey-black fur. In appearance, the shrew can be confused with a European mole, although the tail is shorter and the bare feet more delicate. It likes any habitat providing vegetation cover, particularly marshlands, lush grasslands and damp woodlands and forest, and it is also found around urban gardens and makes nests in cellars and outbuildings. Sight is poor in the short-tailed shrew, so it mainly navigates and finds prey using smell and echolocating clicks. It is a diurnal feeder, living off insects, lizards, frogs, snakes, plants and even voles and mice. Durable plant foodstuffs, such as nuts and seeds, are often stored for later eating. Its saliva is toxic for controlling and killing prey, and the short-tailed shrew also releases an unpleasant musk liquid from scent glands on the body, possibly as an anti-predation measure.

Species name:	*Blarina brevicauda*
Features:	Grey-black fur; bare nose and nose; eyes and ears hidden under fur
Habitat:	Woodlands, forests, marshlands, gardens and agricultural borders
Distribution:	Southern Canada, north and eastern USA
Length:	Up to 14cm (5.5in)
Weight:	Up to 30g (1oz)
Breeding:	Three to 10 pups born after a 21- to 22-day gestation

INSECTIVORA: SORICIDAE

Eurasian Water Shrew

The Eurasian water shrew is equally at home feeding in terrestrial and aquatic environments. Its favoured habitats are forests and also the borders of lakes, ponds, rivers and streams, where it establishes solitary territories. On land, the shrew eats worms, beetles and insects, but it prefers to hunt in the water for fish, frogs, molluscs and aquatic insects (the shrew's saliva is poisonous to weaken the prey). The animal's coat – black on the back and sides, and grey-white underneath – is water-repellent and also traps air to provide buoyancy underwater. Propulsion for the dive and swim comes from the back feet, and the dives can last up to 20 seconds. Water shrews lives in burrows on land, creating nesting chambers full of dry vegetation. They do not hibernate in the winter. Water shrews can be prolific breeders, with a gestation period of only around 20 days and litters averaging five or six, although these can go as high as 12 infants.

Species name:	*Neomys fodiens*
Features:	Black fur on back and sides, grey-white fur underneath
Habitat:	Woodlands, forests and aquatic habitats
Distribution:	Europe to northern Asia
Length:	Up to 9.5cm (3.7in)
Weight:	Up to 25g (0.8oz)
Breeding:	Two to 12 infants born after a 20-day gestation

INSECTIVORA: SORICIDAE

Eurasian Shrew

The Eurasian shrew is common throughout the Eurasian territory, its success owing to its great adaptability to different terrains and local food supplies. It is found mainly in woodlands and grasslands, but also inhabits hedgerows and the borders of aquatic environments. It grows to a maximum body length of around 8cm (3.2in), and its fur is dark brown on the back, lightening through to grey-white on the underparts. 'Opportunistic' is the best description of its feeding habits. It is principally carnivorous, feeding on insects, spiders, worms, woodlice, carrion and anything else it can viably handle. It attacks its prey using a bite powerful for its size. Shrews are solitary and take aggressive control of their home ranges, readily fighting intruders. Males and females create resting nests in sheltered locations, while the females also construct larger breeding nests out of leaves and grasses during the appropriate season. Six or seven young are born here, and these will be weaned after 30 days.

Species name:	*Sorex araneus*
Features:	Dark brown fur on back, paler fur on sides, grey-white fur underneath
Habitat:	Woodlands, forests, grasslands
Distribution:	Europe to northern Asia
Length:	Up to 8cm (3.2in)
Weight:	Up to 14g (0.5oz)
Breeding:	Six to seven infants born after a 19- to 21-day gestation

INSECTIVORA: TALPIDAE

Star-Nose Mole

The star-nosed mole is so-called after its highly unusual snout. This consists of 22 sensory tentacles (known as 'rays') extending out from around the nostrils, which are used to feel for prey and study its terrain (its eyes can sense only the contrast between light and dark). The tentacles move rapidly during foraging, examining up to 12 objects per second (prey identification takes around half a second), but they can also close over the nose to prevent dirt ingress when digging. Apart from the nose, the star-nosed mole has the typical appearance of many other moles – dense black fur and wide, powerful front limbs for digging. Star-nosed moles are semi-aquatic animals, using their strong limbs to swim underwater. Their tunnel systems contain nesting chambers, and when dug into a riverbank, these often feature entrances that are located beneath the waterline. The star-nosed mole, however, lives in a wide variety of habitats, including woodlands, taiga, marshes and bogs – anywhere with plenty of dampness in the soil.

Species name:	*Condylura cristata*
Features:	Dark brown to black fur; star-shaped snout with 22 sensory rays
Habitat:	Woodlands, forests, taiga, rivers, lakes, ponds and marshlands
Distribution:	Eastern Canada and north-eastern United States
Length:	Up to 19cm (7.5in)
Weight:	Up to 45g (1.6oz)
Breeding:	Two to seven infants born after a 45-day gestation

INSECTIVORA: TALPIDAE

Russian Desman

Desmans are endangered members of the mole family. The two principal species are the Russian desman (*Desmana moschata*), which inhabits Eastern Europe and Central Asia, and the Pyrenean desman (*Galemys pyrenaicus*). Unlike moles, desmans are quite at home in aquatic environments – the feet are all webbed and the very long, flattened tail (the same length as the animal's head and body) provides propulsion and steering through the water. The Russian desman averages around 19cm (7.5in) in body length and is covered in a very dense red-brown fur (which fades to a grey on the underparts), and its snout is long, flexible and grooved. This snout is the ideal organ for hunting down prey on the riverbed and in riverbanks – the desman's diet is typically fish, frogs, water insects and crustaceans. Up to eight Russian desmans will live in a single burrow. The main danger facing desmans (both Pyrenean and Russian) is habitat loss to agriculture and pollution, which affect the stocks of animals on which the desman can feed.

Species name:	*Desmana moschata*
Features:	Dark brown-red fur; webbed feet; long, flattened tail; long snout
Habitat:	Freshwater lakes, rivers, streams and ponds
Distribution:	Eastern Europe and Central Asia
Length:	Up to 21cm (8.3in) without tail
Weight:	Up to 450g (16oz)
Breeding:	Three to five infants born after a 40- to 50-day gestation

INSECTIVORA: TALPIDAE

European Mole

The common European mole is found from Britain to the northwestern reaches of Asia. It has a familiar appearance – an almost total covering of short black fur, the only exposed skin being on the tip of the nose and on the paws. The front paws are superb digging tools, being broad with long claws (the rear paws are relatively undeveloped). The mole will spend much of its time underground, and by using its front legs in a rhythmic scooping motion it can excavate up to 20m (66ft) of tunnel every day – molehills are evidence of the digging, and their presence in fields and gardens often attracts the persecution of humans. Within the tunnels, bed chambers are lined with grass and moss for warmth. Earthworms form the bulk of the European mole's diet. However, the mole can take a variety of other (mainly soil-dwelling) creatures, including small snakes and lizards. Moles tend to live alone, except during the breeding season.

Species name:	*Talpa europaea*
Features:	Short black fur; powerful front limbs; hairless nose
Habitat:	Grasslands, woodlands and agricultural areas
Distribution:	Western Europe to northwestern Asia
Length:	Up to 16cm (6.3in) without tail
Weight:	Up to 125g (5oz)
Breeding:	One to seven infants born after a 28-day gestation

INSECTIVORA: TENRECIDAE

Common Tenrec

Common tenrecs are mainly found on Madagascar and the Comoro Islands off the coast of East Africa, although there are introduced populations on Reunion, Mauritius and the Seychelle islands. Typical habitats are woodlands, forests and grasslands – anywhere where there is plenty of vegetation to provide cover and accessible sources of water nearby. In common with the 24 other species of tenrec dotted in and around Africa, the common tenrec has a mixed hair/spiny coat (colour ranges through assorted shades of brown) and a shrewlike face. It is mainly nocturnal in behaviour, although it is also seen to be active around dusk. Its diet is fairly broad – as well as insects, it will take frogs, lizards and mice. Tenrecs are nimble climbers, and can easily ascend trees and rockfaces, and they are also good swimmers. During the winter months, tenrecs hibernate in a specially created burrow 1m (3ft 3in) long, blocking up the entrance with soil.

Species name:	*Tenrec ecaudatus*
Features:	Coat of mixed hair and spines; pointed face
Habitat:	Grasslands, woodlands and forests
Distribution:	Madagascar, Comoro Islands, Reunion, Mauritius, Seychelles
Length:	Up to 39cm (15.4in) without tail
Weight:	Up to 2.5kg (5lb 8oz)
Breeding:	10 to 12 infants born after a 50- to 60-day gestation

LAGOMORPHA: LEPORIDAE

Snowshoe Hare

Snowshore hares are North American mammals that grow to a maximum total length of 52cm (22in) – the females are the larger of the sexes. Coat colour change is dramatic and seasonal. During the summer months, the fur is grey-brown with white underparts and some cream colours around the face and ears, but the entire coat goes pure white in the winter. A thick spread of fur on the feet constitutes the 'snowshoe' of its name. Snowshoe hares tend to be solitary, but the number of home ranges within a given area can be numerous and overlapping. The hares are most active during the twilight and night-time hours, and much of the daylight is spent in sleeping or grooming. Communication between the hares often involves thumbing the hind feet onto the ground, which signals danger. The diet is herbivorous, with the hares often eating their own faeces to extract the maximum nutritional value from the original food.

Species name:	*Lepus americanus*
Features:	Grey-brown fur during summer, pure white in winter;
Habitat:	Open fields, hedgerows, marshland, cedar bogs and coniferous woodland
Distribution:	North America
Length:	Up to 47cm (18.5in)
Weight:	Up to 1.5kg (3lb)
Breeding:	Two to eight leverets born after a 36-day gestation

LAGOMORPHA: LEPORIDAE

Arctic Hare

The Arctic hare has a similar appearance to the snowshoe hare, but its dimensions are more substantial – it can grow up to 66cm (26in) from the nose to rump. During the summer, its coat is blue-grey with light underparts, while in the winter the coat changes to pure white. Black tips to the ears are present regardless of season, and act as a useful identifier. The ears themselves are very short, which reduces heat loss in the subzero climate. Feet are extremely broad to give good mobility across loose and deep snow. The Arctic hare is found from northern Canada up to northern Greenland. Its social behaviour is curious: while it is frequently solitary, it can also be found in a group mass of up to several hundred. The group will coordinate its feeding activities, with some standing guard while others eat, and the members also cluster together to share body warmth on the exposed land.

Species name:	*Lepus arcticus*
Features:	Short black-tipped ears; pure white coat in winter, blue-grey in summer
Habitat:	Arctic tundra
Distribution:	Canada and Greenland
Length:	Up to 66cm (26in) without tail
Weight:	Up to 7kg (15lb)
Breeding:	Four to eight leverets born after a 50-day gestation

LAGOMORPHA: LEPORIDAE

Black-tailed Jackrabbit

The black-tailed jackrabbit (actually of the hare family) is instantly recognizable by its disproportionately large ears – a jackrabbit with a body length of around 58cm (23in) can have ears up to 15cm (6in) long. Such large appendages are a useful aberration. Not only do they maximize radiated heat loss during the summer months, but they also give heightened predator detection. Other features of the black-tailed jackrabbit are a black stripe down the centre of the back and a black rump, the overall coat being a dark grey-brown with light brown to cream underparts. Black-tailed jackrabbits are the largest hare species in North America. In flight they can achieve speeds of up to 56km/h (35mph), interrupting their run with leaps and zigzag changes in direction. They usually live in scraped-out surface depressions or under bushes, rather than in burrows, and they tend to lead solitary lives. They are also intensive herbivorous feeders, eating everything from twigs to cacti, and extracting most of their water from their food.

Species name:	*Lepus californicus*
Features:	Extremely long ears; black rump; dark stripe down back
Habitat:	Arid lands, grasslands and scrub
Distribution:	USA and Mexico
Length:	Up to 63cm (25in) without tail
Weight:	Up to 3.5kg (7lb 11oz)
Breeding:	One to six leverets born after a 41- to 47-day gestation

LAGOMORPHA: LEPORIDAE

European Rabbit

While the largest concentrations of European rabbits are found in Europe itself, they also inhabit regions of Australia, New Zealand, Africa and South America. Once introduced into a system, they can rapidly become a pest because of their reproductive capabilities – a single female can produce 30 young in a year (an average of five kits per litter, six times a year on the basis of a 28 to 33-day gestation period). Kits are weaned after just 28 days. European rabbits have black-brown fur with cream patches around the eyes, throat and shoulders and pale underparts. Usual habitats are open grasslands and deciduous (never coniferous) woodlands and forests, although they are also seen in sandy coastal environments. The rabbits live in large colonies, shelter and nesting provided by warrens dug up to 2m (6ft 6in) into the earth, with an expanded grass-lined nesting chamber at one end. The herbivorous diet includes grasses, cereals and bark, and the damage they wreak on agricultural crops leads to open hunting.

Species name:	*Oryctolagus cuniculus*
Features:	Black-brown fur; cream eye rings; black-tipped ears
Habitat:	Grasslands, woodlands, deciduous forests
Distribution:	Europe, Australia, New Zealand, Africa and South America
Length:	Up to 50cm (20in) without tail
Weight:	Up to 2.5kg (5lb 8oz)
Breeding:	Three to eight kits born after 28- to 33-day gestation

MARSUPIALIA: DASYURIDAE

Quoll

Quoll is the Aboriginal word for 'tiger-cat' – the animals are roughly the size of domestic cats, although they have marsupial characteristics and more elaborate coat colour schemes. There are four species of quoll, concentrated in and around Australia: northern, western, eastern and spotted-tailed. Appearance varies according to species, but the eastern quoll (*Dasyurus viverrinus*) of Tasmania is representative. Growing up to 45cm (18in) in the body, with a tail that is 17–28cm (6.7–11in) long, the eastern quoll has a grey-brown to black fur, the body interspersed with prominent white spots (the tail is without spots). Males are much larger than the females. Quolls are carnivorous, nocturnal animals with an aggressive nature. Their diet is small mammals such as mice, rabbits and rats, insects and carrion, although they will also eat fruit. Habitat for the eastern quoll (and all quolls) includes grasslands, forests, scrub and agricultural areas. Females can produce up to 24 young in a single litter, but the mother has only six teats, which means infant mortality is exceptionally high.

Species name:	*Dayurus viverrinus*
Features:	Grey-brown to black fur; body covered in white spots
Habitat:	Grasslands, woodlands, forests, scrub and agricultural areas
Distribution:	Tasmania, Australia
Length:	Up to 45cm (18in) without tail
Weight:	Up to 2kg (4lb 6oz)
Breeding:	Up to 24 young born after a 21-day gestation

MARSUPIALIA: DASYURIDAE

Tasmanian Devil

The Tasmanian devil is confined to the island of its name. It has a justifiable reputation for being vicious – although it is not territorial, it will fight violently over a kill or if it is threatened. Jaws of immense power enable the Tasmanian devil – the largest of carnivorous marsupials – to eat bones and fur with ease, and to kill large prey such as sheep, wallabies and possums, although smaller prey such as insects and snakes form a significant part of the diet. A large adult can have a body length of 80cm (32in). Its fur is black, with white patches on the rump and chest, and exposed skin around the snout. Apart from mating and communal feeding on a large kill, Tasmanian devils live alone, making dens under bushes, rocks and logs, or in existing animal burrows. April is the main birthing month for the devils, the mother producing two or three young. The young will stay in the mother's pouch for up to four months.

Species name:	*Sarcophilus harrissi*
Features:	Black fur with white patches on rump and chest; exposed snout; stocky physique
Habitat:	Grasslands, woodlands, forests and agricultural areas
Distribution:	Tasmania. Australia
Length:	Up to 80cm (32in) without tail
Weight:	Up to 12kg (26lb)
Breeding:	Two or three young born after a 31-day gestation

MARSUPIALIA: MACROPODIDAE

Tree Kangaroo

The tree kangaroo has an appearance somewhere between a primate and a wallaby, retaining the large hind legs and long balancing tail of the latter, although the tree kangaroo finds hopping on the ground awkward – it is unable to move faster than an average walking pace. Its front legs are also long, however, and all feet have strong claws and leathery pads ideally suited for dexterous tree climbing. Tree kangaroos have been seen to make tree-to-tree leaps of up to 9m (30ft) and tree-to-floor drops of more than 15m (50ft) without sustaining an injury. The diet of tree kangaroos tends to be leaves and fruit, which are collected both directly from the branch and during foraging expeditions onto the ground. Additional elements, however, include birds' eggs, small birds, flowers, grain and bark. Tree kangaroos tend to be nocturnal in habits, although they are sometimes seen sunning themselves atop the tree canopy, and they establish territories of up to 25 hectares (61 acres). The data are for the Bennett's tree kangaroo.

Species name:	*Dendrolagus bennettianus*
Features:	Highly developed back legs; clawed feet; dark-brown fur
Habitat:	Rainforests
Distribution:	Northeastern Queensland, Australia
Length:	Up to 75cm (30in) without tail
Weight:	Up to 13kg (29lb)
Breeding:	One infant per year (limited breeding information)

MARSUPIALIA: MACROPODIDAE

Huon Tree Kangaroo

The Huon tree kangaroo – *Dendrogalus matschiei* – is one of the 10 tree-kangaroo species found in New Guinea and Australia. It lives in forest areas of the Huon peninsula in Papua New Guinea, inhabiting territory 900–1500m (2950–4900ft) above sea level. They grow to 60cm (23.6in) in the body, with a tail roughly the same length. The fur is mahogany-coloured on the back – which also features a dark stripe – and the limbs and tail are golden. The face is yellow with a prominent white patch. Much of their lives is spent asleep, waking to feed on plant material, fruit, carrion, eggs, young birds and insects. One joey is born after a gestation period of up to 45 days. This will then stay in the mother's pouch for up to 300 days. Huon tree kangaroos are in an endangered state. They are hunted for meat and for their skins, and habitat destruction from logging and other industries is rampant in their territories.

Species name:	*Dendrogalus matschiei*
Features:	Golden limbs and tail; brown back fur; white face patch; strong limbs; clawed feet
Habitat:	Tropical forests
Distribution:	New Guinea
Length:	Up to 60cm (23.6in) without tail
Weight:	Up to 6kg (15lb)
Breeding:	One joey born after a 39- to 45-day gestation

MARSUPIALIA: MACROPODIDAE

Eastern Grey Kangaroo

Eastern grey kangaroos are sizeable creatures, with a height of up to 1.8m (5ft 10in) – their powerful tails add another 1.2m (3ft 11in) – and weights of up to 60kg (132lb). While the smaller western grey kangaroo inhabits southern Australia, the eastern grey is found in eastern regions and also in Tasmania. Eastern greys gather together in 'mobs', groups consisting of around three males (one of them being the dominant male) and three females and their young. When moving at speed, the eastern grey has formidable pace, with each hop covering up to 8m (26ft) of distance. The kangaroos are primarily nocturnal, sheltering from the hot Australian sun during the daytime under bushes, trees and rocks. After it is born, the tiny joey crawls through the mother's fur into the pouch to feed off the teats. It lives in the pouch for around 11 months, and is not weaned until around 18 months. Its diet is mainly grasses, leaves and herbs.

Species name:	*Macropus giganteus*
Features:	Brown fur with pale chest and belly; large, angular hind limbs; long thick tail
Habitat:	Grasslands, woodlands, forests and scrub
Distribution:	Eastern Australia including Tasmania
Length:	Up to 1.8m (5ft 10in) without tail
Weight:	Up to 60kg (132lb)
Breeding:	One joey born after a 31-day gestation

MARSUPIALIA: MACROPODIDAE

Whip-tailed Wallaby

The whip-tailed wallaby, also known as Parry's wallaby, grows to over 2.1m (7ft) and is typically identified through the white stripes along the cheeks and throat, and its white underparts and hind legs. The rest of the fur is a pale brown-grey. Whip-tailed wallabies are highly communal, living in groups of up to 80 animals, although within this large group the wallabies tend to form more intimate sub-groups of around 10 creatures. Unlike most wallabies, the whip-tailed wallaby extends its habitats into high-altitude terrain, as long as vegetation cover and food sources remain. The diet consists of leaves, grasses and ferns, and most feeding during the hot summer months is conducted in the early morning or late afternoon. Whip-tailed wallabies are common animals and have few predators, apart from dingoes, but large numbers are killed each year on the roads. A distinctive behavioural trait of wallabies is the licking of the arms during hot weather in an attempt to get cool.

Species name:	*Macropus parryi*
Features:	Pale brown-grey fur; white underparts and lower hind legs; white stripes on cheeks and throat
Habitat:	Open forest in hilly and mountainous terrain
Distribution:	Australia
Length:	Up to 90cm (2ft 11in) without tail
Weight:	Up to 17kg (37lb 8oz)
Breeding:	One joey born after a 31- to 34-day gestation

MARSUPIALIA: MACROPODIDAE

Red-necked Wallaby

Red-necked wallabies have behavioural traits typical of many examples of the wallaby family. They tend to feed around dusk and dawn, they lick themselves to keep cool and they have the same form of motion. Although the majority of the fur is grey, with the underparts creamy, the back of the neck and shoulders has a reddened colour, hence the animal's name. In the past, the red-necked wallaby was much hunted for its fur, and its population was further endangered by persecution from farmers, who not only disliked the animal for its grazing habits that denuded pasture used by cattle and sheep, but who also cleared wallaby habitats for agricultural land. The red-necked wallaby is indeed an intensive grazer, and can associate in groups of up to 30 animals during feeding sessions, although socially it prefers to be solitary. In addition to grass, they will also feed on agricultural crops, so licensed hunting of the wallaby is permitted in Australia. Red-necked wallabies are mainly found in eastern and southeastern Australia.

Species name:	*Macropus rufogriseus*
Features:	Grey fur with cream underparts; red-brown fur on back of neck and shoulders
Habitat:	Open forest and shrubby grasslands
Distribution:	Australia
Length:	Up to 1m (3ft 3in) without tail
Weight:	Up to 18.6kg (41lb)
Breeding:	One joey born after a 30-day gestation

MARSUPIALIA: MACROPODIDAE

Red Kangaroo

The red kangaroo is one of the most recognizable of all marsupials, as well as being the largest species in the *marsupialia* order – it can grow up to 1.6m (5ft 3in) long and weigh as much as 90kg (200lb). The thick tails, which are used for balance and communication, can be as long as 106cm (42in). This powerful creature cannot walk, but is capable of hopping at speeds of up to 48km/h (30mph) and leaping over obstacles up to 3.6m (12ft) high. During battles for social dominance or for mates, the red kangaroos will fight vigorously, scratching, wrestling and pushing with the forelimbs and making extremely potent kicks with its hind legs. If a predator is detected (the red kangaroo has excellent senses of sight and hearing), the kangaroo signals alarm by thumping the tail or a foot on the ground. Red kangaroos live in groups of up to 10 animals called 'mobs'. They are very sensitive to drought, with males and females often becoming infertile when water becomes very scarce, and numbers falling dramatically.

Species name:	*Macropus rufus*
Features:	Males: red fur with white underparts and limbs; females: blue-grey fur; powerful hind legs and tail
Habitat:	Arid open grassland, scrubland and desert regions
Distribution:	Australia
Length:	Up to 1.6m (5ft 3in) without tail
Weight:	Up to 90kg (200lb)
Breeding:	One joey born after a 30- to 40-day gestation

MARSUPIALIA: MACROPODIDAE

Quokka

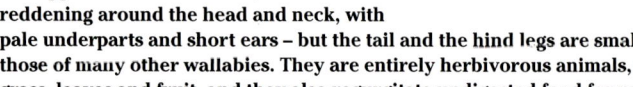

The quokka is a small member of the wallaby family, which inhabits mainly the Rottnest and Bald islands off the southwestern Australian coastline. After approaching extinction, the quokka is now also found in some mainland pockets. (The name 'Rottnest' was given by a Dutch explorer who spotted the quokkas on the island; assuming that they were rats, he named the island 'rat's nest'.) Current population estimates are in the region of 10,000. Quokkas have a typical wallaby-like appearance – brown fur, reddening around the head and neck, with pale underparts and short ears – but the tail and the hind legs are smaller than those of many other wallabies. They are entirely herbivorous animals, feeding off grass, leaves and fruit, and they also regurgitate undigested food for re-chewing. The quokka will be a member of a family group, and this in turn is part of several other groups that make up a large territorial pack. The family group is patriarchal, and it will go out at night on foraging expeditions.

Species name:	*Setniz brachyurus*
Features:	Brown fur, red-brown fur around neck and head areas; pointed snout
Habitat:	Open grassland and scrubland
Distribution:	Rottnest and Bald islands; pockets in southwestern Australia
Length:	Up to 48cm (19in) without tail
Weight:	Up to 4.5kg (10lb)
Breeding:	One joey born after a 27-day gestation

MARSUPIALIA: MACROPODIDAE

Pademelon

There are several species of pademelon, these being wallaby-like creatures which live in rainforests and woodland habitats. A representative example of the pademelons is the red-legged pademelon (*Thylogale stigmatica*), which inhabits forested areas of northern and eastern Australia and similar terrains in New Guinea. Grey-brown to fawn in colour (the lighter-coloured fur tends to be on animals in more open terrain), with white underparts, the red-legged pademelon is a grazing animal, feeding on leaves, bark and fallen fruit. Usually the pademelons live alone, but will feed communally. Within the temporary feeding group, a dominant male will have control, and each animal will have to respect a proprietary distance between the others to avoid fights – 30–50m (98–164ft) is the typical distance of separation. During rest periods, the pademelon will often sit on the base of the tail, with the rest of the tail snaking forward through the hind legs.

Species name:	*Thylogale stigmatica*
Features:	Grey-brown fur; pale underparts; slim head shape
Habitat:	Tropical forests and woodlands
Distribution:	Northern and eastern Australia; New Guinea
Length:	Up to 58cm (23in) without tail
Weight:	Up to 7kg (15lb)
Breeding:	One infant born after a 28- to 30-day gestation

MARSUPIALIA: MYRMECOBIIDAE

Numbat

Numbats are also known as banded anteaters and walpurtis, and they inhabit the southwest corner of Australia. Although the numbat does indeed eat ants, the bulk of its diet is termites – it tears open termite mounds with its forefeet claws, then licks up the termites with its tongue, which is 10cm (4in) long. It feeds voraciously, and an adult numbat can eat up to 20,000 termites in a single day. Numbats have a reddish-brown fur with bands of white colouration crossing the back, and a long, very bushy tail. Suited to its feeding activity, the head is slim and pointed, and features black stripes from the muzzle to the eyes. Numbats are diurnal and solitary, although during the mating season males will move long distances from their home ranges to find breeding females. The young are born in grass- and leaf-lined underground chambers; normal nesting sites are typically hollow logs or abandoned burrows. Numbats are an endangered species, mainly due to increased predation following the introduction of the European fox into Australia.

Species name:	*Myrmecobius fasciatus*
Features:	Reddish-brown fur; white bands across back; pale underparts; slim head shape
Habitat:	Woodlands and grasslands
Distribution:	Southwestern Australia
Length:	Up to 28cm (11in) without tail
Weight:	Up to 700g (1lb 8oz)
Breeding:	Around four infants born after a 14-day gestation

MARSUPIALIA: PERAMELIDAE

Long-nosed Bandicoot

The long-nosed bandicoot is typical of many small, insect-eating marsupials with its elongated snout, which is used for probing the earth and insect nests for food. An adult has a head and body length of between 31 and 43cm (12 and 17in) and lives in forests, woodlands and bushlands around eastern Australia. The long-nosed bandicoot is not quite as imperilled as many other bandicoot species, but decades of hunting by Aborigines (for food), by farmers (to protect crops) and a host of other predators have led to protection orders. When foraging, the long-nosed bandicoot digs distinctive conical-shaped holes in the ground, the right shape for it to explore with its long snout. The bulk of the diet is insects, but the bandicoot will also eat small rodents, roots and tubers. The female of the species has one of the shortest gestation periods in the animal kingdom – only 12.5 days – after which around four young are born, these feeding on the eight teats in the mother's pouch.

Species name:	*Perameles nasuta*
Features:	Grey-brown fur on back and sides; creamy-white fur underneath; long pointed nosed
Habitat:	Woodlands, rainforests and grasslands
Distribution:	Eastern Australia
Length:	Up to 43cm (17in) without tail
Weight:	Up to 1.2kg (2lb 10oz)
Breeding:	Around four infants born after a 12½ day gestation

MARSUPIALIA: PETAURIDAE

Striped Possum

As with many Australian marsupials, the striped possum is indigenous to a relatively confined territory – in this case, northeastern Australia and New Guinea. It shares a similarity with the skunk in terms of appearance – a black-and-white striped body and head, and a long bushy tail, which is dark on top, but lighter underneath with a white tip. It can also release a noxious liquid from the anal glands when threatened. Striped possums spend most of their time in the trees and are exceptionally agile climbers, observed swinging between branches and sleeping in tree forks. They hunt for wood-boring insects, extracting these from the wood by using the sharp front teeth (which they apply to strip away the bark), a long tongue and an extra-long fourth finger on the front paws. Other elements of the striped possum diet include various types of insect, birds, rodents and fruit. Feeding and other activities usually take place at night.

Species name:	*Dactylopsila trivirgata*
Features:	Black-and-white striped body; white-tipped long bushy tail
Habitat:	Rainforests
Distribution:	Northeastern Australia and New Guinea
Length:	Up to 28cm (11in) without tail
Weight:	Up to 500g (1lb)
Breeding:	One or two infants (gestation not known)

MARSUPIALIA: PETAURIDAE

Sugar Glider

Sugar gliders have a membrane stretching between the front and rear limbs that, once outstretched, turns the marsupial into a virtual hang-glider. Stretching this out and leaping from trees, the sugar glider can achieve flights of up to 45m (148ft) in distance. With the membrane closed, the animal has a fairly diminutive appearance, with a body length of up to 32cm (12.6in) and a tail (free from the flight membrane) of up to 48cm (19in). Overall body colour is blue-grey, with a prominent dark stripe running the length of the spine from the forehead. Black stripes run across each eye, and the rest of the face is white. Sugar gliders are socially quite structured, forming mixed-sex groups of up to seven adults plus young, with territories (an individual territory is around 1 hectare (2.5 acres) based around a certain area of trees in which they live and feed. The territory is marked out using saliva and scent-gland emissions. Sugar gliders eat a highly varied diet, including eucalyptus tree sap, pollen, insects and small mammals.

Species name:	*Petaurus breviceps*
Features:	Blue-grey body; black stripe from forehead to tail; black stripes across eyes
Habitat:	Forests
Distribution:	Australia, (including Tasmania), New Guinea, parts of Indonesia
Length:	Up to 15cm (6in) without tail
Weight:	Up to 171g (160oz)
Breeding:	One or two infants after a 16-day gestation

MARSUPIALIA: PETAURIDAE

Ring-tailed Possum

The ring-tailed possum (also known as the common ringtail) has grey-brown fur (reddish-brown when young) with creamy underparts, while the tail is split between dark and then white colouration. It is an arboreal resident of eastern mainland Australia and Tasmania, its tree-climbing aided by two thumbs on each foot, which give it a superb grip on branches, and a highly prehensile tail that is used for carrying vegetation for nesting, as well as providing balance. Habitat range is broad, and extends from large forests and woodlands through to parks and gardens within urban areas. Ring-tailed possums are principally nocturnal, spending the day in one of several nests (known as 'drays') built within a territory. The drays are generally thick balls of grass and leaves constructed in tree hollows or patches of dense vegetation, but in urban areas the ring-tailed possum is also found nesting in house roofs. The ringtail diets mainly on leaves, and joins species such as the koala in being able to eat the toxic eucalyptus leaves.

Species name:	*Pseudocheirus peregrinus*
Features:	Grey-brown body; white underparts; black-and-white tail
Habitat:	Arboreal environments; urban parkland
Distribution:	Eastern Australia (including Tasmania)
Length:	Up to 35cm (14in) without tail
Weight:	Up to 1.1g (39oz)
Breeding:	One to four infants (usually two) after a 30-day gestation

MARSUPIALIA: PHALANGERIDAE

Cuscus

There are five species of cuscus concentrated in various parts of mainland Australia, Tasmania, New Guinea, the Solomon Islands, the Moluccas and Celebes. They are tree-dwelling marsupials of diverse size – they range from 15 to 66cm (6–26in) – and feed on a diet of fruit, leaves, insects and occasionally small vertebrates. General characteristics of the cuscus include opposable toes on the hind feet, a strong prehensile tail for gripping and balancing, and a thick woolly coat. The heads have a flattened appearance. Behaviourally, the cuscus species are usually nocturnal, and move in a very deliberate but confident manner through the trees. A typical cuscus species is the common cuscus (*Phalanger orientalis*), also known as the grey cuscus. Its fur colour is highly variable, but it is distinguished by a dark stripe along the centre of the face and spine and a hairless tail section. The common cuscus inhabits tropical rainforests and other areas of dense vegetation, and lives out a solitary life, although many are kept as pets on account of their friendly nature.

Species name:	*Phalanger orientalis*
Features:	White to grey-brown body; black stripe along face and spine; tail bare at end
Habitat:	Tropical rainforest
Distribution:	Southwest Pacific
Length:	Up to 48cm (19in) without tail
Weight:	Up to 3.5kg (7lb 11oz)
Breeding:	One to three infants after a 13-day gestation

MARSUPIALIA: PHALANGERIDAE

Brush-tailed Possum

The brush-tailed possum is distributed throughout much of mainland Australia and Tasmania, living in a variety of habitats according to regional geography. Most of these habitats are arboreal in nature – rainforests, mangrove forests, woodlands, parks and gardens – but in some more arid areas they can live much of their lives on the ground. Nesting places include tree holes, hollowed-out logs, termite mounds and rock crevices, while urban habitations include outbuildings and attics. Appearance varies according to the location, but generally the possum has a silver-grey coat of short, dense fur, with paler underparts. The feet have sharp claws to aid climbing, and the rear feet have opposable first toes. Brush-tailed possums are nocturnal and solitary (although home ranges often overlap because of high population densities), and they live mainly on plant foods, including the leaves of poisonous plants such as eucalyptus (birds' eggs are sometimes eaten). The possums breed about twice a year, the infant remaining in the mother's pouch for about four months.

Species name:	*Trichosurus vulpecula*
Features:	Variable coat colour (red to dark grey); bushy prehensile tail; large A-shaped ears
Habitat:	Arboreal environments and scrubland; rural housing
Distribution:	Australia (including Tasmania)
Length:	Up to 58cm (23in) without tail
Weight:	Up to 4.5kg (10lb)
Breeding:	One infant after a 18-day gestation

MARSUPIALIA: PHASCOLARCTIDAE

Koala

The koala is often erroneously labelled the koala 'bear' on account of its appearance, particularly its rounded, tufted ears and black muzzle, and the thick fur, which provides not only insulation, but also a high degree of water resistance. Koalas are arboreal creatures. They do not have a significant tail for balance; instead the koala relies on strong limbs and powerful clawed hands, which have rough pads and, in the forelegs, an opposable grip (two fingers, opposed to three). The koala survives on the leaves of the eucalyptus tree, a diet that is nutritionally barren and poisonous to most other animals. It feeds during the night, but sleeps between 18 and 22 hours a day, wedging itself in the fork of a tree. Socially the koala lives in groups, with each individual establishing a home range that usually overlaps with those of others. Males may also be solitary. Koala numbers are precarious – habitat loss to agriculture and urbanization is a major threat, and thousands are killed each year by domestic dogs and cars.

Species name:	*Phascolarctos cinereus*
Features:	Thick grey to grey-brown fur; black muzzle, tufted ears strong limbs; clawed feet
Habitat:	Woodlands and forests
Distribution:	East Australia
Length:	Up to 82cm (32in) without tail
Weight:	Up to 15kg (33lb)
Breeding:	One joey born after a 35-day gestation

MARSUPIALIA: THYLACOMYIDAE

Bilby

The bilby is a curious-looking inhabitant of scrublands and grasslands in western and central Australia. It is also called the rabbit-eared bandicoot on account of extremely long ears, and its long back legs also have a rabbit-like quality. The snout is slim and extended, ideally suited for snuffling-up insects and larvae from the ground (the bilby first excavates the prey from the ground by using its strong forelegs). As well as insects, bilbies will eat some plant foods. Fur colour is blue-grey, while the tail is divided into black and white halves. Bilbies are burrowing animals, and they create their dens by digging passageways about 1–2m (3ft 3in–6ft 6in) deep into the earth, these being served by one or two entrances. Currently the bilby has a 'vulnerable' classification with the IUCN. Its existence is threatened for a number of reasons, including hunting for its fur, habitat destruction through its lands being stripped by livestock grazing, and predation – foxes, dingoes, dogs and cats all exact a heavy toll.

Species name:	*Macrotis lagotis*
Features:	Blue-grey fur; very large ears; thin snout; black-and-white tail
Habitat:	Arid regions and grasslands
Distribution:	Western and Central Australia
Length:	Up to 55cm (22in) without tail
Weight:	Up to 2.5kg (5lb 8oz)
Breeding:	One infant born after a 13- to 16-day gestation

MARSUPIALIA: VOMBATIDAE

Common Wombat

Growing up to 1.2m (3ft 11in) in length, the common wombat is a familiar though ecologically vulnerable resident of eastern mainland Australia and Tasmania. In appearance it is similar to a small bear – it has a rounded, solid body covered in grey-brown to black fur, and short, stocky legs with broad clawed feet, and its head features a naked nose and very short ears. The wombat's claws make it an accomplished digger. It will excavate an extensive burrow network into the slopes of valleys, creeks and gullies, the burrow going down to a depth of 2m (6ft 6in) and branching out into multiple passageways and chambers. On average, the burrows have a total length of around 20m (65ft 7in), but sometimes exceed 100m (328ft). The typical wombat habitat is scrubland, open forest and hilly/mountainous areas. It is a nocturnal, solitary and territorial animal, and is a ferocious fighter when cornered. The diet is purely herbivorous, the preference being for grasses, roots, sedges and herbs. The wombat can create extensive disruption to farmland on account of its diet and burrows, and hence it has in the past been poisoned and shot in huge numbers.

Species name:	*Vombatus ursinus*
Features:	Short grey-brown to black fur; stocky physique; large claws; short ears
Habitat:	Scrubland, open forest, hilly and mountainous terrain
Distribution:	Eastern Australia (including Tasmania)
Length:	Up to 1.2m (3ft 11in) without tail
Weight:	Up to 40kg (88lb)
Breeding:	One infant born after a 30-day gestation

MONOTREMATA: ORNITHORHYNCIDAE

Platypus

There are only five species within the *Monotremata* family, the key distinguishing characteristic being that they are egg-laying mammals. The platypus is an aquatic animal, with broad webbed feet, a flat paddle-like tail, a waterproof coat (this consists of a waterproof outer coat and a dense, insulating inner fur) and a long flat bill. Each hind foot has a sharp spur on the inner side, through which poison can be injected, the usual targets being predators or males competing for breeding rights. Platypuses are territorial creatures, and they live solitary lives. Preferred habitats are rivers, lakes and streams, especially those with gentle currents and overhanging vegetation. The platypus will dig burrows into the banks, and these can measure up to 9m (29ft 6in) in length. Here the female will lay one to three eggs, which hatch after only 12–14 days, then the young are suckled for up to three months. Its diet is a broad range of small aquatic insects and animals, and the platypus's bill is electro-sensitive to help it locate prey through electro-emissions in the water.

Species name:	*Ornithorhynchus anatinus*
Features:	Waterproof deep-brown fur; webbed feet; flat bill and broad, flat tail
Habitat:	Rivers, lakes, streams and some coastal areas
Distribution:	Eastern Australia (including Tasmania)
Length:	Up to 60cm (23.6in) without tail
Weight:	Up to 2.5kg (5lb 8oz)
Breeding:	One to three eggs laid 12–14 days after mating

MONOTREMATA: TACHYGLOSSIDAE

Short-nosed Echidna

The short-nosed echidna is the smaller relative to the long-nosed echidna (*Zaglossus bartoni*); unlike the endangered long-nosed animal, which is confined purely to New Guinea, the short-nosed echidna is fairly common across mainland Australia, Tasmania and New Guinea. It is also called the spiny anteater, and its body is covered with yellow-black (occasionally pure yellow) spines that extend up to 6cm (2.4in), beneath which is dark brown to black fur. The snout is very pointed, and seems to account for most of an almost neckless head, and the paws are flat with heavy claws, ideal for digging. Like the platypus, the echidna has a defensive spine on the hind legs, although this is not venomous. The bulk of the echidna diet is ants and termites, the snout being used to probe into logs and through foliage, and the claws breaking open nests. The females lay a solitary egg after mating, this egg going directly into the animal's pouch, where it hatches after a period of 9–27 days (the infant has a shell tooth to break its way out).

Species name:	*Tachyglossus aculeatus*
Features:	Long yellow-black spines; pointed snout; large claws on feet
Habitat:	Forests, woodlands, arid regions, grasslands
Distribution:	Australia (including Tasmania) and New Guinea
Length:	Up to 45cm (18in) without tail
Weight:	Up to 7kg (15lb)
Breeding:	One egg hatching after a 9- to 27-day gestation

PERISSODACTYLA: EQUIDAE

African Wild Ass

Recognizably related to the common domesticated ass, the African wild ass survives in some of the most arid zones of East Africa, although it is critically endangered. In appearance it has a grey coat with paler muzzle, throat and legs – some subspecies have legs striped like those of a zebra – and a short, prominent mane tipped with darker hairs. A long tail terminates in a black brush. The animal is mainly seen active at dawn and dusk, avoiding the worst daytime temperatures and grazing on a very wide variety of grasses, leaves and bark. These provide the ass with much of its water needs, although it will need to drink at least once every three days. Socially the African wild ass gathers in fluid mixed-sex herds of up to 50 animals (usually much smaller), and male territories can extend up to 29 sq. km (11 sq. miles). These territories are marked with large piles of dung. African wild asses are agile and fast animals, with a running speed of up to 50km/h (31mph).

Species name:	*Equus africanus*
Features:	Grey coat; paler legs (some species striped legs); upright mane
Habitat:	Arid regions
Distribution:	East Africa
Length:	Up to 2.3m (7ft 6in) without tail
Weight:	Up to 230kg (510lb)
Breeding:	One foal hatching after a 11- to 12-month gestation

PERISSODACTYLA: EQUIDAE

Plains Zebra

The plains zebra, also known as the Burchell's zebra, is the familiar plains-dwelling zebra species of East and southern Africa. It has horse-like proportions, being 2.5m (8ft 3in) long – not counting the bushy black tail – but what defines the zebra is its black-and-white striped body. The stripes generally wrap completely beneath the zebra's abdomen, in contrast to some other zebra species, although in certain regions the stripes will thin out or even disappear on the belly and legs. Like most equids, zebras are highly social – the social unit consists of a dominant stallion and several females plus offspring, although stallions will also form bachelor groups. Females are also arranged into a hierarchy, the stallion's first mate occupying the top spot. Zebras are powerful creatures; males engage in violent kicking and biting fights for breeding rights, and even large predators such as lions take care around the zebra's rear legs. Zebra foals are able to stand on these legs within 15 minutes of being born.

Species name:	*Equus burchelli*
Features:	Black-and-white striped body; high-standing mane, black muzzle and tail bush
Habitat:	Open savannah
Distribution:	East and southern Africa
Length:	Up to 2.5m (8ft 3in) without tail
Weight:	Up to 260kg (570lb)
Breeding:	One foal hatching after a 370-day gestation

PERISSODACTYLA: EQUIDAE

Mountain Zebra

The mountain zebra is found in scattered areas of southern Africa, and both of its subspecies – *Equus zebra zebra* and *Equus zebra hartmannae* – are classified as endangered by CITES. The striped patterns on the mountain zebra widen on the haunches and run up through the mane. A dewlap under the throat also separates the mountain zebra from the plains zebra. As the name suggests, the mountain zebra is at home in rocky and mountainous terrain, being a nimble climber. The diet in this terrain is typical of zebras – grasses, leaves and herbs – and most feeding activity is conducted in the early morning or late afternoon. Female mountain zebras come to sexual maturity when they are around two years old. They produce one foal after a 12-month gestation, the foal being weaned for the first 10 months of its life within the herd. Herds consists of a dominant stallion, around six females and the offspring, and the home ranges extend up to 20 sq. km (7.7 sq. miles). Mountain zebras are frequently seen taking dust baths, these aiding parasite control.

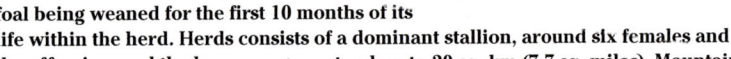

Species name:	*Equus zebra*
Features:	Black-and-white striped body; high-standing mane, black muzzle and tail bush
Habitat:	Deserts, semi-deserts, rocky and mountainous terrain
Distribution:	Southern Africa
Length:	Up to 2.2m (7ft 3in) without tail
Weight:	Up to 370kg (814lb)
Breeding:	One foal born after a 370-day gestation

PERISSODACTYLA: RHINOCEROTIDAE

White Rhinoceros

The white rhinoceros is one of five rhino species, and it lives in scattered and frequently endangered pockets throughout sub-Saharan African. A fully grown adult male can total 2.3 tons (2.5 tonnes) in weight and reach up to 4m (13ft) from rump to nose. The rhino's body has massive solidity, with a skin colour ranging through greys and browns. Distinguishing features are skin folds at the top of the legs, a prominent nuchal crest atop the neck (a muscular concentration to provide head support), and two horns on the snout, which are made of keratin fibres instead of horn. The anterior horn is the longer, sometimes exceeding 1m (3ft 3in) in length. White rhinos are grass-eaters, the straight upper lip allowing for intensive grazing. While males frequently live alone, mothers and offspring can form herds of up to seven animals. Territories are marked by urine spraying. The population of white rhinos is highly precarious, and extinction is a real threat, particularly among the northern white rhinoceros subspecies (*Ceratotherium simum cottoni*).

Species name:	*Ceratotherium simum*
Features:	Grey–brown skin; skin folds at top of legs; nuchal crest; horned snout
Habitat:	Savannah
Distribution:	Pockets in West, East and southern Africa
Length:	Up to 4m (13ft) without tail
Weight:	Up to 2.3 tons (2.5 tonnes)
Breeding:	One calf born after a 16-month gestation

PERISSODACTYLA: RHINOCEROTIDAE

Sumatran Rhinoceros

The Sumatran rhinoceros has the unfortunate title of being one of the world's rarest animals, its population classified as critically endangered owing to overhunting and habitat loss. A recent study in the Malaya Peninsula placed the population density at one animal per 40 sq. km (15 sq. miles). Its habitats range from rainforests to marshlands, but it seems to prefer densely forested country on hilly or mountainous slopes (it is an able climber), where it diets on grasses, leaves, shoots, shrubs, saplings, fruit and twigs. Sumatran rhinos need access to plenty of water – they are also good swimmers – and each territory will contain a salt lick. Territories are held by females only, the males having solitary and nomadic habits. Feeding activity is usually confined to the night, and much of the day is spent mud wallowing to cool the skin. The mud also serves to shield the skin from insects. Sumatran rhinos are also known as hairy rhinos; the body features shaggy hair in varying concentrations.

Species name:	*Dicerorhinus sumatrensis*
Features:	Grey-brown skin; skin fold across shoulders; shaggy, sporadic hair covering; two rounded horns on snout
Habitat:	Tropical rainforest, swamplands
Distribution:	South and Southeast Asia
Length:	Up to 3.2m (10ft) without tail
Weight:	Up to 950kg (2000lb)
Breeding:	One calf born after a 15- to 16-month gestation

PERISSODACTYLA: RHINOCEROTIDAE

Black Rhinoceros

The black rhinoceros is distinguished from the white rhinoceros by its rounded snout and pointed, prehensile upper lip. This structure is designed for stripping leaves off branches – the black rhino is a grazer and feeds mainly on leaves and branches taken from shrubs and trees. Atop the snout are the two horns, the anterior horn being the longer and growing up to 1.4m (4ft 6in) in length. Black rhinos are grey like most other rhino species, the name coming from the colour of local mud when dried onto the rhino's skin. Black rhino numbers have been decimated by hunting, freefalling from around 65,000 in 1970 to an estimated 2500 today. Poachers target rhinos for their horns, which are used in traditional Chinese medicines and which are more valuable, weight for weight, than gold. If left alone, a black rhino can live for 30 years in the wild. They are solitary, apart from a mother with young offspring, and territorial, marking out their territory with dung piles and urine.

Species name:	*Diceros bicornis*
Features:	Grey skin; two horns on snout; pointed upper lip
Habitat:	Grasslands and savannah
Distribution:	Pockets in sub-Saharan Africa
Length:	Up to 3.2m (10ft) without tail
Weight:	Up to 1.3 tons (1.4 tonnes)
Breeding:	One calf born after a 15- to 16-month gestation

PERISSODACTYLA: RHINOCEROTIDAE

Indian Rhinoceros

The Indian rhinoceros has heavy skin folds, which look like plating rather than skin. The skin on the flanks and rump features close-set tubercles (lumpy structures), and the snout has a single long horn (hence the Indian rhino's scientific name, *Rhinocerus unicornis*). The Indian rhino is one of the biggest of the rhino species, growing up to 3.8m (12ft) in body length. Today it is confined to protected areas of Nepal and Assam, India. It lives in tall grasslands and well-watered forest areas, feeding mainly off grasses, stripping off the tips with a prehensile top lip. It is solitary and territorial, but will associate at mud sites and drinking areas. Fights occur mainly during the breeding season, and the males inflict injuries using their horns and sharp incisors. As with all rhinos, Indian rhinos have excellent hearing and smell to make up for poor eyesight. They can also swim well and run at speeds up to 56km/h (35mph).

Species name:	*Rhinocerus unicornis*
Features:	Grey skin with heavy skin folds; single horns on snout; pointed upper lip
Habitat:	Grasslands and forests
Distribution:	Nepal and India
Length:	Up 3.8m (12ft) without tail
Weight:	Up to 2.2 tons (2.4 tonnes)
Breeding:	One calf born after a 15- to 16-month gestation

PERISSODACTYLA: TAPIRIDAE

Malayan Tapir

The Malayan tapir lives in Southeast Asia, and it has an instantly recognizable skin pattern – the front half of the body and the legs are black; the rear half of the torso is white. (Note that infant Malayan tapirs have a brown coat with white spots and stripes, the adult coat appearing at around seven months of age.) Tapirs are woodland and forest dwellers, and they are ideally shaped for pushing through dense forest undergrowth. Their long snouts contain large nasal passages terminating in a very flexible nose; tapirs have a superb sense of smell, backed by acute hearing. The tapir's feet have three splayed toes, which provide good traction on the softest, muddiest forest ground. Malayan tapirs are mainly nocturnal, with a sure-footed handling of steep slopes and difficult forest terrain. They use well-worn feeding paths, and their home ranges (marked out with urine) are within easy reach of water. Their diet is leaves and shoots. Leopards and tigers are a big threat to the Malayan tapir but humans pose a greater danger.

Species name:	*Tapirus indicus*
Features:	Black head, legs and shoulders, white saddle; elongated, trunklike nose
Habitat:	Forests
Distribution:	Southeast Asia
Length:	Up to 2.5m (8ft 2in) without tail
Weight:	Up to 540kg (1190lb)
Breeding:	One infant born after a 390- to 403-day gestation

PERISSODACTYLA: TAPIRIDAE

South American Tapir

Also known as the Brazilian tapir, *Tapirus terrestris* is found throughout northern and central South America. It grows to a shoulder height of around 91cm (36in) and has a thick, bristly brown coat, lightening to a paler brown on the face, throat and chest. A slim mane also runs down the back of the neck. As with most tapirs, the South American tapir is at home around aquatic habitats, and is a good swimmer. Such are their skills in the water that South American tapirs are even able to dive and feed off the river bottom; being able to dive also makes a useful escape route from jaguars and pumas, the tapir's main animal predators. Brazilian tapirs are usually active during the night hours, feeding on fruit, leaves, ferns and other plant foods. They are generally solitary, but will gather in some numbers around a drinking area or a salt lick. Young tapirs may also stay with their mothers for up to two years.

Species name:	*Tapirus terrestris*
Features:	Bristly brown coat; short mane down neck; elongated, trunklike nose
Habitat:	Forests and inland aquatic environments
Distribution:	Northern and central South America
Length:	Up to 2m (6ft 6in) without tail
Weight:	Up to 250kg (550lb)
Breeding:	One infant born after a 13-month gestation

PHOLIDOTA: MANIDAE

Temminck's Pangolin

Pangolins are also more descriptively called scaly anteaters – their bodies are covered with thick, overlapping plates from head to tail. The scales are brown to yellow in colour, and are made of a hardened form of hair. When it is threatened, the pangolin rolls up and the plates act as a form of armour (they also provide camouflage). The pangolin has a low, long profile similar to that of an armadillo, with a body length of up to 60cm (23.6in) and a tail around 10cm (4in) shorter. There are seven species of pangolins, and they live mainly off ants and termites, ripping the nests open with their strong front claws. The insects are licked up with a long tongue, this being lubricated by a very productive salivary gland to make it sticky. The prey is not chewed (the pangolin does not have teeth), but instead is mashed up by the stomach muscles. Pangolins are vulnerable to human hunting, its meat apparently tasting similar to that of duck.

Species name:	*Manis temmincki*
Features:	Body covered in yellow-brown scales; short legs with heavy claws; slim face
Habitat:	Grasslands, forests and woodlands
Distribution:	Eastern and southern Africa
Length:	Up to 60cm (23.6in) without tail
Weight:	Up to 18kg (40lb)
Breeding:	One or two infants born after a 4-month gestation

PINNIPEDIA: ODOBENIDAE

Walrus

The walrus has an enormous bulk – a large adult male can weigh up to 2 tons (2.2 tonnes) and measure up to 3.5m (11ft 5in). (The males weigh up to 50 per cent more than females). The skin colour is grey-brown, but when the animal is sunbathing the colour reddens significantly, and it becomes paler underwater as blood flow to the skin is reduced. Both males and females have large canine tusks, extending up to 1m (3ft 3in) in some cases. The males use their tusks for sparring during the mating season (January to March), and serious injuries can occur (males often carry scars). The males attract mates by bellowing, and the female will give birth to one calf after a gestation period of 10–11 months, the pregnancy including up to four months of delayed implantation. Walruses are very social, and groups numbering in the thousands inhabit ice floes and coastlines. The sexes separate outside of the breeding season.

Species name:	*Odobenus rosmarus*
Features:	Rough skin and coarse hair; skin colour changeable; large tusks
Habitat:	Arctic aquatic
Distribution:	All Arctic waters
Length:	Up to 3.5m (11ft 5in)
Weight:	Up to 2 tons (2.2 tonnes)
Breeding:	One calf born after a 10- to 11-month gestation (following period of delayed implantation)

PINNIPEDIA: OTARIIDAE

Southern Fur Seal

Southern fur seal is a generic name for nine species of seal that inhabit waters in the southern hemisphere (*see* Northern fur seal). Their territorial range is wide, inhabiting coastlines and offshore waters as far apart as Australia, South America and Antarctica, and consequently there are differences in species size. The Galapagos fur seal grows to around 1.5m (5ft) in length, whereas the Cape fur seal of southern Africa is up to 1m (3ft 3in) longer. Typically, southern fur seals have grey-brown to black coats, although colour variations are common, especially around the neck and shoulder areas. Fur seals take their name from the acute density of their fur, the lush inner coat forming the seal's principal insulating layer (it is more important than the blubber). The fur is so effective at retaining heat that the seals will often crawl out onto rocks and use their flippers as fans to cool themselves. The southern fur seal diet is mainly fish, squid and octopus. The data below are for the Cape fur seal.

Species name:	*Arctocephalus pusillus*
Features:	Dense grey-brown to black fur; dark flippers; prominent ear flaps
Habitat:	Coastal and offshore waters
Distribution:	Southern African, Australia (including Tasmania)
Length:	Up to 2.3m (7ft 6in)
Weight:	Up to 360kg (790lb)
Breeding:	One pup born after a 12-month pregnancy

PINNIPEDIA: OTARIIDAE

Northern Fur Seal

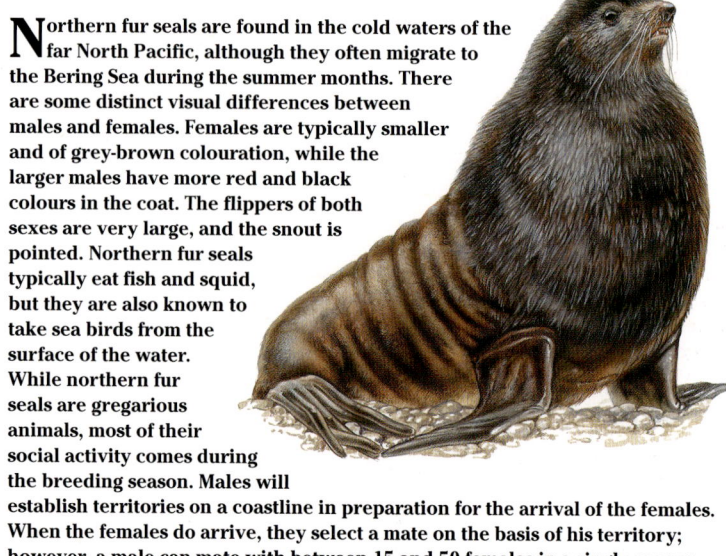

Northern fur seals are found in the cold waters of the far North Pacific, although they often migrate to the Bering Sea during the summer months. There are some distinct visual differences between males and females. Females are typically smaller and of grey-brown colouration, while the larger males have more red and black colours in the coat. The flippers of both sexes are very large, and the snout is pointed. Northern fur seals typically eat fish and squid, but they are also known to take sea birds from the surface of the water. While northern fur seals are gregarious animals, most of their social activity comes during the breeding season. Males will establish territories on a coastline in preparation for the arrival of the females. When the females do arrive, they select a mate on the basis of his territory; however, a male can mate with between 15 and 50 females in a single season. The newborn pups are quite helpless, and the mother will leave them for several days while she goes out to collect food.

Species name:	*Callorhinus ursinus*
Features:	Grey-brown to reddish fur depending on sex; large flippers;
Habitat:	Coastal and offshore waters
Distribution:	North Pacific
Length:	Up to 2.1m (7ft)
Weight:	Up to 270kg (600lb)
Breeding:	One pup born after a 12-month pregnancy (including period of delayed implantation)

PINNIPEDIA: OTARIIDAE

California Sea Lion

Californian sea lions inhabit the coastal waters of the western United States and the Galapagos Islands. They are extremely sociable and inquisitive creatures, and enjoy entering human aquatic environments such as harbours and marinas in the search for food, even coming out of the water to bask on jetties and oil rigs. For this reason, they are also popular attractions in aquatic theme parks, and are used as research animals for the US military. Adult male California sea lions have dark-brown coats, while female colouration is lighter; both have prominent foreheads. Like most seals, the Californian sea lion is a consummate swimmer and diver, with recorded dive depths of 274m (890ft) and swim speeds of up to 32km/h (20mph). As feeders, the sea lions are either social or solitary, depending on the amounts of food available – if dealing with a large shoal of fish, they might feed in a group. The mouths of freshwater rivers are popular feeding places, and everything from squid to guillemots is taken.

Species name:	*Zalophus californiacus*
Features:	Dark brown to tan fur; prominent forehead
Habitat:	Coastal waters
Distribution:	Western USA, Galapagos Islands
Length:	Up to 2.4m (7ft 9in)
Weight:	Up to 390kg (860lb)
Breeding:	One pup born after an 11-month pregnancy (including period of delayed implantation)

PINNIPEDIA: PHOCIDAE

Hooded Seal

The hooded seal is an unusual-looking creature with an inflatable nasal chamber forming a trunklike structure over the mouth. The chamber can be inflated to dramatically expand the size of the head, an effect used during mating (in an attempt to achieve dominance) or when threatened by predators. With about 6.3 litres (13 pints) of capacity, the chamber almost doubles the size of the seal's head. Skin colour is black on the head, with the body being red and covered with white 'cloud' markings. The seals have clawed front flippers. Hooded seals inhabit waters of the North Atlantic (as far down as New England) and up to the Arctic. They diet on fish (especially herring, cod, halibut, turbot and redfish), as well as shrimp, mussels, octopus and squid. Pups are born to females after a 12-month gestation, following which the animals are weaned in only four to five days. The pups are bluish in colour for the first year of life.

Species name:	*Cystophora cristata*
Features:	Black fur on head, brown with white patterning on body; inflatable trunklike nasal chamber
Habitat:	Ice packs and deep waters
Distribution:	North Atlantic and Arctic
Length:	Up to 2.7m (8ft 10in)
Weight:	Up to 410kg (900lb)
Breeding:	One pup born after a 12-month pregnancy

PINNIPEDIA: PHOCIDAE

Grey Seal

The grey seal is found in the North Atlantic and the Baltic, with some large populations around the Canadian and northern European coastlines. It is still the victim of large-scale hunting in places, although European seals have enjoyed increased protection since the mid-1990s. Grey seals have a grey-brown to tan colouration with silver (male) or black (female) spotting. Males and females also have contrasting nose silhouettes, the male's nose being more Roman-shaped and longer. Grey seals breed in 'rookeries' along coastlines and ice floes, taking over beaches, caves and any other feature that provides a suitable site. Males will enter the female breeding sites and mate with up to 10 females. Following birth, the infant is nursed by its mother for two to three weeks (the mother will often not eat during this period). The mother will then leave the pup, which survives on its blubber until it is able to enter the sea to feed.

Species name:	*Halichoerus grypus*
Features:	Grey-brown to tan fur with contrasting silver or black markings; males: pronounced Roman nose
Habitat:	Coastlines, ice environments, deep offshore waters
Distribution:	North Atlantic and the Baltic
Length:	Up to 2.5m (8ft 3in)
Weight:	Up to 310kg (680lb)
Breeding:	One pup born after an 11-month pregnancy

PINNIPEDIA: PHOCIDAE

Leopard Seal

The leopard seal is large (up to 3.2m/10ft long) and is an aggressive member of the seal family, known for its developed hunting skills. Leopard seal diets are broader than those of many other seals; not only will it eat fish and krill, but it also attacks and kills other seals (particularly crabeater and fur seals), penguins, sea birds and carrion. Its killing is done with canine teeth that grow to 2.5cm (1in) in length. In general appearance, the leopard seal is streamlined, with a powerful head and broad shoulders. Colouration is silver or grey, darkening on top, with silver or grey spots contrasting with the background. Leopard seals are usually solitary in behaviour, loitering around pack-ice areas where they will find the greatest concentrations of their prey types. It is a common species, but human krill fishing could adversely affect numbers, as krill forms around 45 per cent of the leopard seal diet.

Species name:	*Hydrurga carcinophagus*
Features:	Grey to silver body with contrasting spots; lithe, muscular body
Habitat:	Antarctic and sub-Antarctic waters
Distribution:	Across southern hemisphere within habitat type
Length:	Up to 3.2m (10ft)
Weight:	Up to 455kg (1000lb)
Breeding:	One pup born after an 11-month pregnancy (including period of delayed implantation)

PINNIPEDIA: PHOCIDAE

Weddell Seal

Weddell seals are usually seen on flat ice fields around Antarctic and sub-Antarctic water. They will lie in the weak southern sun for hours, soaking up what meagre warmth there is by turning onto their back, sides and stomach, and positioning themselves perpendicular to the sun to maximize exposure. The ice is also used as a grooming surface. A typical adult measures around 2.6m (8ft 6in) long, and has a coat that is silver-grey on the back, darkening underneath, with mottled patterning and a black head. Weddell seals are formidable divers, with recorded dives being tracked down to 600m (1968ft) and lasting up to one hour. Fish and squid are central to the diet, which the Weddell seal will hunt beneath the pack ice. The seal has long incisor teeth, and uses them to bite through thin ice to create breathing holes. Their eyes are specially adapted to work in the low-light conditions beneath ice fields.

Species name:	*Leptonychotes weddelli*
Features:	Silver-grey to black body with contrasting spots; black head and tail
Habitat:	Antarctic and sub-Antarctic waters
Distribution:	Across southern hemisphere within habitat type
Length:	Up to 2.9m (9ft 6in)
Weight:	Up to 600kg (1320lb)
Breeding:	One pup born after a 9- to 10-month gestation (after period of delayed implantation)

PINNIPEDIA: PHOCIDAE

Elephant Seal

Elephant seals are of awesome scale – the largest examples of southern elephant seals (*Mirounga leonina*) grow up to 6m (20ft) in length, with a weight of up to 5 tons (5.5 tonnes). There are two species, the southern elephant seal (the larger of the two) is found in Antarctic and sub-Antarctic waters, while the northern elephant seal (*Mirounga angustirostris*) dwells in Pacific coastal waters up to the Gulf of Alaska. The males of both species are distinguished by a large inflatable nose. This is inflated during the breeding season, and the seal will also roar in an attempt to establish dominance. Fighting also occurs during this time, consisting of slapping and butting. For both males and females, the coat colour is grey to dark brown. Fish, octopus and squid form the elephant seal's diet, and it can dive down to 600m (1968ft) while hunting, remaining submerged for up to 40 minutes. Elephant seals are also migratory creatures, travelling several thousand miles between breeding and seasonal habitats. The data below are for the southern elephant seal.

Species name:	*Mirounga leonina*
Features:	Grey to brown body; males: trunklike inflatable nose
Habitat:	Antarctic and sub-Antarctic waters
Distribution:	Across southern hemisphere within habitat type
Length:	Up to 6m (20ft)
Weight:	Up to 5 tons (5.5 tonnes)
Breeding:	One pup born after an 11-month pregnancy (including period of delayed implantation)

PINNIPEDIA: PHOCIDAE

Monk Seal

There have been three species of monk seal – the Mediterranean (*Monachus monachus*), the Hawaiian (*Monachus schauinslandi*) and the Caribbean (*Monachus tropicalis*). The Caribbean monk seal was finally declared extinct in 1996, having been hunted to destruction by hunters who appreciated the monk seal's curiosity. The remaining two species are critically endangered. The Mediterranean monk seal is typical of the genus, although information is scarce about the creature, owing to depleted numbers and a reclusive nature. A male monk seal grows to around 2.5m (8ft 2in) and has brown or grey fur on the back that lightens under the belly. (Pups are black with a white or yellow patch on the belly.) Mediterranean monk seals generally inhabit sea caves around the Mediterranean coastline and the eastern Atlantic, but their sensitivity to pollution, environment disturbance and disease makes for a precarious future, with the population possibly not exceeding 500 animals.

Species name:	*Monachus monachus*
Features:	Grey to brown body, lighter underneath
Habitat:	Temperate coastal waters
Distribution:	Mediterranean and eastern Atlantic
Length:	Up to 2.8m (9ft 2in)
Weight:	Up to 400kg (880lb)
Breeding:	One pup born after a 12-month pregnancy

PINNIPEDIA: PHOCIDAE

Harp Seal

While many seal species have found their populations at risk, the harp seal has managed to remain common within its broad North Atlantic and Arctic Ocean territories, with the population numbering anywhere up to four million animals. (However, both the exhaustion of food stocks and increased hunting may see a contraction of population numbers over the next decade.) Harp seals are migratory animals, travelling 4000km (2500 miles) from the Arctic to the northern coastline of Canada during the autumn, and returning in the summer. They live on ice sheets, which are essential for the seals' breeding and moulting (hence global warming is a threat). They live alone outside of the breeding season, but tens of thousands congregate during the breeding season, although a male and female pair will be monogamous for the season. Harp seals are silver-white in colour, with black eye patches and black harp-shaped markings on the back. Female markings are paler.

Species name:	*Pagophilus groenlandicus*
Features:	Silver-white body; black eye patches, black curved marking on back
Habitat:	Ice floes, coastal and offshore waters
Distribution:	North Atlantic and Arctic Ocean
Length:	Up to 1.7m (5ft 6in)
Weight:	Up to 130kg (290lb)
Breeding:	One pup born after an 11½-month pregnancy (including period of delayed implantation)

PINNIPEDIA: PHOCIDAE

Common Seal

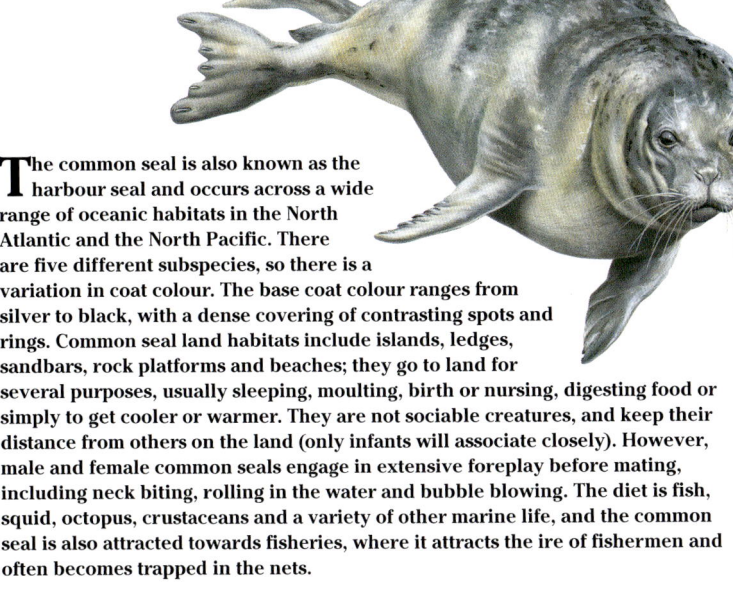

The common seal is also known as the harbour seal and occurs across a wide range of oceanic habitats in the North Atlantic and the North Pacific. There are five different subspecies, so there is a variation in coat colour. The base coat colour ranges from silver to black, with a dense covering of contrasting spots and rings. Common seal land habitats include islands, ledges, sandbars, rock platforms and beaches; they go to land for several purposes, usually sleeping, moulting, birth or nursing, digesting food or simply to get cooler or warmer. They are not sociable creatures, and keep their distance from others on the land (only infants will associate closely). However, male and female common seals engage in extensive foreplay before mating, including neck biting, rolling in the water and bubble blowing. The diet is fish, squid, octopus, crustaceans and a variety of other marine life, and the common seal is also attracted towards fisheries, where it attracts the ire of fishermen and often becomes trapped in the nets.

Species name:	*Phoca vitulina*
Features:	Silver to black bodies with contrasting spots and stripes
Habitat:	Coastal/deep waters' one species inhabits freshwater environments
Distribution:	North Atlantic and North Pacific
Length:	Up to 1.9m (6ft 3in)
Weight:	Up to 170kg (370lb)
Breeding:	One pup born after a 9- to 11-month pregnancy (including period of delayed implantation)

PRIMATES: CALLITRICHIDAE

Common Marmoset

Common marmosets are forest-dwelling residents of southeastern Brazil. They have white ear tufts (and hence are also known as the white-tufted-ear marmoset) and a white forehead blaze, and the body fur is dark to greyish brown. Marmosets are extremely sociable primates, and they live together in groups numbering from two–13. They work together as a group to look after the young (adolescent animals are often appointed as helpers), even if the young are not directly related, and provide the infants with grooming and food. Female common marmosets usually give birth to twin offspring, and the male will stay with the family unit after birth, helping to carry the young. (Note, however that females will mate with multiple males.) The common marmoset has a broad diet within the rainforest habitat. Tree sap forms a large and important part of the diet, but fruit, flowers, nectar, lizards, insects, frogs and eggs are also eaten when necessary.

Species name:	*Callithrix jacchus*
Features:	Grey-brown fur; white ear tufts and blaze on forehead; tail stripes
Habitat:	Forest environments
Distribution:	Southeastern Brazil
Length:	Up to 18cm (7in) without tail
Weight:	Up to 250g (8oz)
Breeding:	Twin offspring after a 148-day gestation

PRIMATES: CALLITRICHIDAE

Pygmy Marmoset

Pygmy marmosets are the world's smallest monkey (not the smallest primate) – an adult marmoset may grow to only 12cm (7in) in length and weigh 100g (3½oz). They are found in the Upper Amazon basin and are arboreal creatures, living only in well-watered forest environments. They live, eat, feed and mate within the trees, and form groups (known as 'troupes') of a male/female pair and their offspring, the total group usually numbering from five–10. Occasionally another male may be present, but one of the males will assert dominance. The pygmy marmoset has a speckled tawny fur with some yellow and green highlights (the tip of the hair can be a different colour to the base), and the tail features dark rings. Despite its size, the pygmy marmoset moves rapidly through the trees, gripping with the sharp claws on its fingers and toes. The marmoset eats a wide range of plant foods, insects and small animals, but its main food is tree gum, which it extracts by stripping away the bark with its teeth and claws.

Species name:	*Callithrix pygmaea*
Features:	Speckled tawny fur with some yellow and green colouration; ringed tail
Habitat:	Forests (especially by rivers) and agricultural fields
Distribution:	Western South America
Length:	Up to 15cm (6in) without tail
Weight:	Up to 125g (4oz)
Breeding:	Two offspring (occasionally three) after a 119- to 140-day gestation

PRIMATES: CALLITRICHIDAE

Golden Lion Tamarin

The golden lion tamarin is known for its luscious long red-gold coat, which flows outwards like a mane over the shoulders, framing the bare grey face. It is on the IUCN's critically endangered list, as the remaining 1500 wild animals live in a tiny area of eastern Brazil that has been massively deforested over the past 40 years; thankfully the Brazilian government now watches over the tamarins in a protected territory. Golden lion tamarins inhabit rainforest and swamp-forest environments, where they mainly survive on fruit, but also eat gum, nectar, insects and lizards. They are diurnal animals, feeding during the daytime and spending the night in a tree hole. As with many other marmoset and tamarin species, golden lion tamarins live in groups consisting of a breeding pair plus offspring, typically numbering two to nine animals. The group is internally caring, sharing food with the young and all adults carrying infants. Territory sizes are around 40 hectares (100 acres), marked out with scent markings on the trees.

Species name:	*Leotopithecus rosalia*
Features:	Bushy golden-red fur; long clawed toes; grey face
Habitat:	Forests
Distribution:	East coast of Brazil
Length:	Up to 25cm (10in) without tail
Weight:	Up to 800g (29oz)
Breeding:	Two offspring (occasionally three) after a 126- to 130-day gestation

PRIMATES: CALLITRICHIDAE

Emperor Tamarin

The emperor tamarin is defined by the long, white moustache that grows out from the face; the name is ascribed to the resemblance between the tamarin and the moustached German emperor Wilhelm II. The rest of the body is grey-brown and the underparts are white, while the tail is a rusty orange colour. Emperor tamarins are found in parts of Brazil and Peru in South America, the preferred habitat being forests environments, both lowland and montane. The typical emperor tamarin social group has two to eight creatures, with a dominant female leading several males and the offspring (the female is polyandrous, meaning she will mate with all the males in the group). All members of the group will contribute towards the welfare of the young. Interestingly, the emperor tamarin will often associate with tamarins of different species, such as the brown-mantled tamarin. The emperor tamarin's omnivorous diet includes fruit, sap, insects, frogs and birds' eggs.

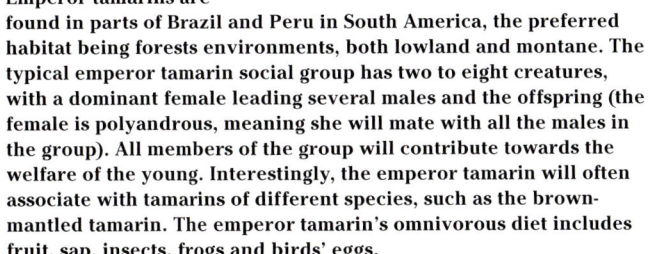

Species name:	*Saguinus imperator*
Features:	Long white moustache; grey-brown body; orange tail
Habitat:	Forests
Distribution:	Western South America
Length:	Up to 25cm (10in) without tail
Weight:	Up to 450g (16oz)
Breeding:	Two offspring (occasionally three) after a 140- to 145-day gestation

PRIMATES: CALLITRICHIDAE

Cotton-Top Tamarin

Cotton-top tamarins are small, striking animals with a black face framed by an elaborate white crown of fur. Its back and tail are black-brown while the underparts are white. They are found in a small region of Colombia, but the species is now endangered owing to the use of these creatures as pets and as medical subjects (the animals were used in experiments for bowel-cancer treatments). Their habitats are deciduous forests and rainforests, and they tend to stay in the trees, briefly going to ground to forage for food among fallen vegetation. Cotton-top tamarins eat the diet typical of their family – fruit, sap, insects, small vertebrates – which they collect during the daylight hours, travelling distances of around 2km (1.2 miles) to find food supplies (the tamarin will have to travel further during the food scarcities of the dry seasons). Most of the foraging is conducted at the mid-levels of the trees. Sleep is taken in the fork of a branch or amidst a patch of dense vegetation.

Species name:	*Saguinus oedipus*
Features:	Bushy white head crown; black face; black-brown body and tail; white underparts
Habitat:	Tropical and deciduous forests
Distribution:	Colombia
Length:	Up to 25cm (10in) without tail
Weight:	Up to 450g (16oz)
Breeding:	One or two offspring after a 125- to 140-day gestation

PRIMATES: CEBIDAE

Red Howler Monkey

Red howler monkeys are aptly named – their powerful howl (the loudest noise of any land animal), delivered from high in the treetops, can be heard over 2km (1.2 miles) away. The howl serves to communicate with neighbouring groups of monkeys, and tends to be used in the mornings and when the group transfers to new feeding grounds. By alerting other groups to their presence, the howler monkeys can avoid potential clashes, although they do not have fixed or defended territories. Red howler monkeys have fur of various shades of red, often with a lighter saddle. The face is especially deep, and the tail is long and prehensile. The typical group numbers four–11 animals, governed by a single dominant male. They are diurnal creatures, although they will spend about 50 per cent of the day sleeping and resting. Infants will remain with the group until they are independent, then leave. This behaviour is probably precautionary – intruding adult males may kill the offspring to bring the mother into season.

Species name:	*Alouatta seniculus*
Features:	Deep face; red body and tail
Habitat:	Tropical, swampy and deciduous forests
Distribution:	Northwest South America
Length:	Up to 63cm (25in) without tail
Weight:	Up to 9kg (20oz)
Breeding:	One offspring (occasionally two) after a 180- to 194-day gestation

PRIMATES: CEBIDAE

Red Uakari

The red uakari, falling under the species name bald uakari, takes both its names from its red/pink hairless face. The colour of the rest of the body depends on the subspecies, and ranges from pure white through to the red uakari's orange. Tails are short and bushy. Red uakaris are found in the northwest of South America, straddling the Colombia–Brazil borderlands. They are an endangered species, their numbers threatened by hunting and by deforestation (their rainforest homelands are being decimated by logging). Highly gregarious by nature, the uakari will gather in mixed-sex groups of 10–20 animals, with the maximum number approaching 100. They also associate with other monkey species when feeding. Although red uakaris sometimes eat insects, their diet is principally herbivorous, mainly fruit, nectar and leaves. The dry season brings the animal increasingly down to the forest floor to forage for seeds and other fallen plant foods.

Species name:	*Cacajao calvus*
Features:	Bald red face; orange fur; tail relatively short; clawless feet
Habitat:	Tropical forests
Distribution:	Northwest South America
Length:	Up to 57cm (22.4in) without tail
Weight:	Up to 3.5kg (7lb 12oz)
Breeding:	One offspring (occasionally two) after unknown gestation period

PRIMATES: CEBIDAE

Brown Capuchin

The brown capuchin (also known as the black-capped monkey) is a diurnal animal that is, like almost all other South American primates, highly sociable. The typical group numbers up to 14 animals, although it can climb as high as 40, and this group will contain a mix of males, females and offspring. A dominant male and dominant female lead the group, and these two members tend to keep themselves separate from the rest. The group will forage en masse, spreading out through the trees or across the forest floor and covering a radial distance of about 100m (328ft). Infants find a safe dynamic within the group, and are often seen playing vigorously together during the evening hours. Brown capuchins are found in tropical and deciduous forests throughout much of South America, although deforestation is shrinking their habitat. The typical diet is fruit, nuts and small vertebrates, and capuchins will often use stones and sticks to open resistant food.

Species name:	*Cebus apella*
Features:	Light to dark brown coat; dark-tipped prehensile tail; black 'cap' of fur on the head
Habitat:	Tropical and deciduous forests
Distribution:	South America
Length:	Up to 42cm (17in) without tail
Weight:	Up to 4.5kg (10lb)
Breeding:	One offspring after a 153- to 161-day gestation

PRIMATES: CEBIDAE

Woolly Monkey

Woolly monkeys, of which there are four subspecies, are found in the Amazonian forests of central South America. They are aptly named, having a very thick short coat that ranges from grey-brown to yellow, which provides superb insulation in some of the monkey's high-altitude montane forest habitats. A woolly monkey colony has from 5–50 members, and this establishes a territorial home range of about 4 sq. km (1.5 sq. miles). However, woolly monkeys are not aggressively territorial, and will often permit monkeys from outside the territory to enter and feed. Woolly monkey colonies are patriarchal, being headed by a dominant male and several other high-ranking males. All males involve themselves with care of the young, often carrying the infants on their backs. Woolly monkeys are arboreal creatures, with powerful limbs, opposable toes on the feet and a strong prehensile tail that has a hairless grip pad just beneath the tip.

Species name:	*Lagothrix lagotricha*
Features:	Grey-brown to yellowish coat; prehensile tail with gripping pad; powerful shoulders and large, square head shape
Habitat:	Tropical rainforests
Distribution:	South America (Amazon basin)
Length:	Up to 65cm (26in) without tail
Weight:	Up to 10kg (22lb)
Breeding:	One offspring after a 223-day gestation

PRIMATES: CEBIDAE

Squirrel Monkey

Squirrel monkeys are residents to Costa Rica and Panama, their habitats being primary and secondary forests, and they are also seen around agricultural areas. They are almost entirely arboreal in behaviour, rarely venturing down to the forest floor. Squirrel monkeys have a grey coat and yellowish legs, and their black-tipped tails are longer than their heads and bodies – they will often wrap the tail over a shoulder when they are resting. The head is capped with black fur, and their hands have nails rather than claws. Squirrel monkeys are inquisitive and playful, and so they have suffered the attentions of tourists, who often disturb their habitats. A squirrel monkey colony usually numbers around 30–50 animals, but several hundred have been recorded. There are dominant males and females within the colony, and the colony also separates into sub-groups arranged by sex and social position. The males will fight for breeding rights during the September to November mating season.

Species name:	*Saimiri sciureus*
Features:	Grey coat; yellowish legs; very long prehensile tail; black fur cap
Habitat:	Arboreal environments
Distribution:	South America
Length:	Up to 32cm (12?in) without tail
Weight:	Up to 1kg (2lb 3oz)
Breeding:	One offspring after a 160- to 170-day gestation

PRIMATES: CERCOPITHECIDAE

Diana Monkey

The Diana monkey has become an endangered species thanks to the familiar causes of environmental destruction in Africa – hunting and deforestation. They have also been used extensively in medical research. They are arboreal and diurnal creatures which associate together in groups of 15–30 individuals, dominated by an adult male. The daily routine is extremely active, with the young practising their climbing skills in nearby trees and the older members heading off to forage for leaves, fruit, flowers, insects, small vertebrates and any other accessible foods. Much time is also spent grooming, and this particularly helps to bond the older members of the group with the infants. Like most primates, the Diana monkey is highly vocal, its vocal noises ranging from clicks to loud shrieks. Its appearance is varied, but typically it has a black-brown coat reddening on the back and thighs, with a white chest and pointed beard.

Species name:	*Cercopithecus diana*
Features:	Black-brown coat; white chest and beard
Habitat:	Forest environments
Distribution:	West Africa
Length:	Up to 59cm (23in) without tail
Weight:	Up to 8kg (18lb)
Breeding:	One offspring (rarely two) after a five-month gestation

PRIMATES: CERCOPITHECIDAE

Colobus Monkey

Colobus monkeys inhabit forest environments in Central and eastern Africa, ranging from Cameroon to Ethiopia. Being reclusive animals, they are often heard rather than seen, their distinctive vocalizations including a rhythmic rasping noise and a loud roar. They are also extremely agile, jumping distances of 6m (20ft) from tree to tree and gripping the branches with their four-fingered hands (colobus monkeys have no thumb). They have black-and-white bodies, white featuring at the tip of the bushy tail, in the long shoulder hairs and as a frame to the face. Their diet is mainly leaves, fruit and flowers, and the colobus monkey's digestive system is specially structured to maximize the nutrition gained from plant foods. Colobus monkeys are aggressively territorial, and the troupe consists of a single male and four or five females with offspring. Even during the breeding season, however, fighting is rare between competing males.

Species name:	*Colobus guereza*
Features:	Mixed black-and-white coat; long shoulder hairs, white face frame
Habitat:	Forest and environments
Distribution:	Central and eastern Africa
Length:	Up to 57cm (22.4in) without tail
Weight:	Up to 13.5kg (30lb)
Breeding:	One offspring after a five- to seven-month gestation

PRIMATES: CERCOPITHECIDAE

Japanese Macaque

Japanese macaques are native purely to the country of their name, although numbers were introduced into Texas in the early 1970s, where they now form a wild group. Their appearance is dominated by a striking red face and bottom, the rest of the coat being plain brown to grey fur. They are forest-dwelling animals, and will inhabit a broad range of forest environments, deciduous to coniferous, and are mainly found in mountainous territories on Honshu, Japan's principal island. Coats on the Japanese macaque are very thick to enable them to endure the freezing Japanese winters in the mountains, although they are observed bathing in heated volcanic springs to keep warm. The macaque group is of mixed sex, but the females will outnumber the males by a factor of three. Males tend to move in and out of the group, while females will establish hierarchy. Interestingly, an infant will inherit its mother's place within the hierarchy when it reaches adulthood.

Species name:	*Macaca fuscata*
Features:	Red face and bottom; brown to grey coat; short tail
Habitat:	Montane forest environments
Distribution:	Japan; introduced population in Texas
Length:	Up to 95cm (37in) without tail
Weight:	Up to 14kg (31lb)
Breeding:	One offspring after a 173-day gestation

PRIMATES: CERCOPITHECIDAE

Rhesus Monkey

Rhesus monkeys inhabit many habitats in Central, South, East and Southeast Asia (principally Afghanistan, India, Thailand and southern China), but with a population shrinking under the effects of human interference. They have been extensively used in medical and psychological research, and the discovery of rhesus antigens in their blood in the 1940s led to the advanced classification of human blood groups. Habitat types for the rhesus monkey extend from wooded flatlands to montane forests up to 3000m (9842ft) in elevation, and they can survive in temperatures ranging from the tropical to polar. They have a brown fur that often lightens on the back and thighs, and both face and rump are a reddish colour. Their social organization is complex, with males being the dominant animals, but the females making up the practical leadership of the group. Group numbers have been known to exceed 100 animals, although beyond this point splinter groups will usually be formed.

Species name:	*Macaca mulatta*
Features:	Red face and bottom; brown coat
Habitat:	Forests, woodlands, montane forests, flatlands
Distribution:	Asia
Length:	Up to 64cm (25in) without tail
Weight:	Up to 12kg (26lb 8oz)
Breeding:	One offspring after a 135- to 194-day gestation

PRIMATES: CERCOPITHECIDAE

Lion-tailed Macaque

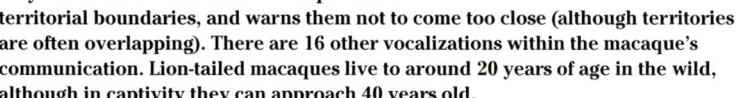

The lion-tailed macaque (other name 'wanderoo') is critically endangered, with current research estimating only 400 animals left in the wild. They live in southwestern India, in the Ghats mountains, and their habitats are mountainous rainforests and monsoon forests up to an altitude of 1500m (4921ft). The rarity of the species affects social composition, but the typical family group numbers from 10–20 individuals, although this can climb above 30. Within the group there are one to three males, only one of these being dominant (and therefore having the right to breed). The males will establish a group territory using vocal communication only – a loud howl alerts other troupes to the territorial boundaries, and warns them not to come too close (although territories are often overlapping). There are 16 other vocalizations within the macaque's communication. Lion-tailed macaques live to around 20 years of age in the wild, although in captivity they can approach 40 years old.

Species name:	*Macaca silenus*
Features:	Shaggy brown-grey mane around face; black body and tail; tail features tufted tip
Habitat:	Montane forests
Distribution:	Southwestern India
Length:	Up to 61cm (24in) without tail
Weight:	Up to 10kg (22lb)
Breeding:	One offspring after a 162- to 186-day gestation

PRIMATES: CERCOPITHECIDAE

Barbary Ape

Barbary apes, more accurately called Barbary macaques, are resident to North Africa and also the Mediterranean island of Gibraltar. They have grey coats varying with shades of yellow and brown, and a dark pink face. Distinctively for a monkey species, they also have no tail. Barbary apes are mainly herbivorous, their diet including fruit, leaves, roots, nuts, seeds and acorns, but they will also eat selected invertebrates such as caterpillars. They have extremely large cheek pouches, and when in a hurry the apes can stuff these with food equivalent to their stomach volume. In North Africa, the principal habitats for the ape are cedar, oak and scrub forests up to altitudes of 2000m (6561ft). They live in matrilineal hierarchies, the females leading the group and passing on status to offspring. Females will breed with multiple males, and the uncertainty about paternity leads the males to look after any young within the group.

Species name:	*Macaca sylvanus*
Features:	Grey-brown to yellow-brown coats; pink face; no tail
Habitat:	Cedar, oak and scrub forests
Distribution:	Algeria, Morocco, Tunisia and Gibraltar
Length:	Up to 76cm (30in) without tail
Weight:	Up to 13kg (29lb)
Breeding:	One offspring after a 196-day gestation

PRIMATES: CERCOPITHECIDAE

Mandrill

Mandrills are instantly identified through their spectacular face colouration. The nose is deep red, the colour running from the nostrils in a strip up the bridge of the nose to the eyes. Either side of the nose the muzzle is made up of blue ridges, and a yellow beard runs beneath the chin. The bulk of the body is olive brown with white underparts, and the rump is blue to purple. Male madrills have much more intense colours than the visually muted females. Mandrills are extremely powerful creatures, with opposable thumbs for gripping and extremely large canine teeth – these can be more than 6cm (2.4in) long. The mandrill society is strictly patriarchal; each group will have a dominant male with exclusive breeding rights over the females, who will usually outnumber the males substantially. Multiple groups of mandrills may also inhabit the same territory to form troupes of over 200 animals. Mating tends to occur about once every two years rather than on a seasonal basis.

Species name:	*Mandrillus sphinx*
Features:	Dramatic blue-and-red face colouration; yellow beard; blue to purple rump; olive-grey body fur
Habitat:	Rainforest, secondary and montane forests
Distribution:	Western Africa
Length:	Up to 81cm (32in) without tail
Weight:	Up to 37kg (82lb)
Breeding:	One offspring after a 220- to 270-day gestation

PRIMATES: CERCOPITHECIDAE

Proboscis Monkey

Proboscis monkeys are found purely on Borneo, where they inhabit rainforests, swamplands, mangrove forests and coastal regions. All territories are located with water no more than several hundred metres away. Both males and females have large noses, the male's unusually so – it hangs down over the mouth like a pendulum, which is thought to increase the power of vocalizations. The proboscis monkey is a diurnal and arboreal creature. It is extremely nimble in the trees, having an easy quadrupedal movement through the tree canopy. In addition, proboscis monkeys are good swimmers, and will often leap from the trees into water below. Socially, proboscis monkeys split into either male-only groups or groups with a single dominant male ruling a number of females and offspring. Typically the group size ranges from three–30 animals. In uni-male groups, male offspring are cared for within the group until they are around 18 months of age, after which they are forced out.

Species name:	*Nasalis larvatus*
Features:	Unusually large nose; partially webbed feet; pink-brown body fur, red on the head and shoulders
Habitat:	Rainforests, swamplands, mangrove forests, coastal regions
Distribution:	Borneo
Length:	Up to 76cm (30in) without tail
Weight:	Up to 21kg (46lb)
Breeding:	One offspring after a 166-day gestation

PRIMATES: CERCOPITHECIDAE

Hamadryas Baboon

The hamadryas baboon has a similar appearance to the African baboon, and is found in Ethiopia, Somalia, Saudi Arabia and the Yemen. Naturalists have been especially interested in this baboon's highly complex social organization. A group of up to 10 baboons is the basic social block, which consists of one dominant male, females plus offspring. Several groups are organized into a band, and several bands make a troupe. The male hierarchy is patrilineal, with sons taking over the father's status and inheriting his females. Furthermore, males will aggressively acquire females from non-familial bands, but will use more communicative methods with females from their own bands. Home ranges can extend over territory of 40 sq. km (15.4 sq. miles). Hamadryas baboons are omnivorous animals, and will eat leaves, fruit, grasses, gun seeds, flowers, roots and bulbs, eggs and small vertebrates. Their habitats are arid, scrubby regions, and consequently the wide variety in their diet is essential to their survival.

Species name:	*Papio hamadryas*
Features:	Long brown fur; males have silver shoulders and back; hairless face and rump
Habitat:	Arid lands
Distribution:	North/East Africa and Middle East
Length:	Up to 76cm (30in) without tail
Weight:	Up to 21.5kg (47lb)
Breeding:	One offspring after a 172-day gestation

PRIMATES: CERCOPITHECIDAE

Douc Langur

Douc langurs are indigenous to Vietnam, and include three species types: the red-shanked douc langur (*Pygathrix nemaeus nemaeus*), the black-shanked (*Pygathrix nemaeus nigripes*) and the grey-shanked (*Pygathrix nemaeus cinerea*). The names are partially descriptive of the key differences between the species, the red-shanked animal having maroon colouration on the back legs and white forearms, and the black-shanked douc langur sporting black hind legs and grey forearms. *Pygathrix nemaeus cinerea* is mainly grey and black, although like the other subspecies it has touches of orange around the sides of the face. In all the subspecies, the main fur colour is grey, and they also have prominent whisker-like white hairs radiating out from around the face and white tails. Douc langurs mainly live in tropical rainforests and monsoon forests, where they form groups of four–50 animals. While males are dominant over the group, females also have a hierarchical structure.

Species name:	*Pygathrix nemaeus*
Features:	White tail; maroon section of back legs; black thighs; grey body; white whiskers
Habitat:	Tropical rainforests and monsoon forests
Distribution:	Vietnam
Length:	Up to 76cm (30in) without tail
Weight:	Up to 11kg (24lb)
Breeding:	One offspring (occasionally two) after a 165- to 190-day gestation

PRIMATES: CERCOPITHECIDAE

Hanuman Langur

The Hanuman langur bears the name of the Hindu monkey god Hanuman. It is found throughout the Indian subcontinent, where many communities regard it as a sacred animal – by consequence the langurs benefit from food offerings left out by local people, and are tolerated even when they raid crops. Other countries host to the langur are Bangladesh, Burma and Sri Lanka. In colour their fur ranges from brown to golden, varying according to subspecies and age, and the face and hands are usually darker. They are quadrupedal animals, and most of their time is spent on the ground rather than up in the trees. The diet is the usual mix of plant materials typically eaten by primates, but Hanuman langurs will also eat quantities of soil to give them the correct mineral intake. Social groupings are sometimes uni-male/multi-female or multi-male/multi-female, and range in number from around 10–60. A single male is dominant, and if he has just taken over the group he will often kill existing offspring to stamp out competing genes and bring the females into season.

Species name:	*Semnopithecus entellus*
Features:	Brown to golden fur; black face; dark, long tail
Habitat:	Forest, woodland, arid scrub and urban areas
Distribution:	South Asia
Length:	Up to 78cm (31in) without tail
Weight:	Up to 20kg (44lb)
Breeding:	One offspring (occasionally two) after a 168- to 200-day gestation

PRIMATES: CERCOPITHECIDAE

Gelada Baboon

Gelada baboons have two striking physical features – a long cape of dark brown to tan fur cresting out from the head and flowing down from the shoulders, and a vivid red hourglass pattern on the neck and chest, framed by paler skin. The pink colour is thought to display the animal's virility. Gelada baboons are found purely in the montane grasslands of Ethiopia and Eritrea in East Africa. The grasses provide dietary sustenance, with the grass blades or the grass seeds being the main foods, depending on the season (the presence of grass blades is reduced during the dry season). Socially the gelada baboons have complex structures, with large troupes of up to 300 creatures composed of various bands, in which the social structure varies from multi-male to uni-male. When an outside male attempts to oust a dominant male from his position, the females in the group may work together to repel the intruder.

Species name:	*Theropithecus gelada*
Features:	Long cape of fur; fur mantle around face; red hourglass-shape skin on face
Habitat:	Montane grasslands
Distribution:	East Africa
Length:	Up to 74cm (29in) without tail
Weight:	Up to 19kg (41lb)
Breeding:	One offspring (occasionally two) after a 150- to 180-day gestation

PRIMATES: DAUBENTONIIDAE

Aye-Aye

The aye-aye is an endangered resident of the forestlands and mangrove swamps of northwest and eastern Madagascar. Aye-ayes are one of a number of animals known as 'promisians' (which include lemurs and lorises): these are arboreal creatures with highly developed hands and, usually, exceptional powers of night vision owing to reflective properties in the eyes, which act as natural image intensifiers. Aye-ayes are unusual-looking promisians, with enormous ears, a wide fleshy face, a thick covering of long, coarse black-and-white fur and long skeletal fingers. The middle finger is particularly pronounced, and the aye-aye uses this to probe wood-boring insects out of bark once it has detected them with its sensitive ears and stripped off the outer bark with its sharp teeth. As well as insects, they will also eat fruit, fungi and sap. Aye-ayes teeter on the edge of extinction – as well as deforestation, they are also killed by local people, who view the aye-aye as a symbol of death.

Species name:	*Daubentonia madagascariensis*
Features:	Long middle finger; large ears; long black-and-white coat
Habitat:	Forests, mangroves, bamboo thickets
Distribution:	Madagascar
Length:	Up to 40cm (16in) without tail
Weight:	Up to 3kg (6lb 8oz)
Breeding:	One offspring after a 172-day gestation

PRIMATES: HYLOBATIDAE

Lar Gibbon

The lar gibbon is also known as the white-handed gibbon on account of the white fur around the hands, and also the face (the face itself is black) and the feet. The rest of the body is thickly covered with a fur of uniform colour, which ranges from a pale cream to dark-brown. Lar gibbons are arboreal masters, living high up in the tree canopy of Southeast Asian rainforests. They move at speed through the canopy via a process called 'brachiation', swinging from branch to branch and using their body weight to provide pendulum-like momentum to leap across gaps of several metres with a acute judgment of distance. The fingers are specially developed for this movement; when swinging, they are formed into a flexible hook-like structure. Much remains to understand of the lar gibbon's social world. A common social grouping consists of a monogamous male/female pair plus offspring, although the adults will occasionally make new partnerships. Males and females make extremely loud vocal duets, which serve to establish their territorial boundaries against outsiders.

Species name:	*Hylobates lar*
Features:	White fur around hands, face (which is black) and feet; thick sitting pads on rump
Habitat:	Rainforests
Distribution:	Southeast Asia
Length:	Up to 59cm (23in) without tail
Weight:	Up to 7.5kg (17lb)
Breeding:	One or two offspring after a seven-month gestation

253

PRIMATES: HYLOBATIDAE

Siamang Gibbon

The siamang is the largest member of the gibbons. Its presence is heightened by an arm spread of up to 1.5m (ft), the same as the animal's standing height. The siamang is covered in a thick black fur and is tailless. A distinctive feature is its elastic throat sac, which swells outwards dramatically to amplify the siamang's shrieks and other calls (it can reach the same proportions as a human head). Male and female siamangs form monogamous pairs, and the two will perform extremely loud vocal duets in order to ward off the incursions of other males into the home range and to establish the boundaries of the territory. They are particularly vocal in the morning just before the day's activity begins. When moving through the trees, the siamangs use the hand-over-hand brachiation action classic to gibbons. Such is the strength of the arms that the siamang will often hang on one hand while it feeds. The siamang diet is principally leaves; indeed, leaves constitute more of the diet than seen in any other gibbon.

Species name:	*Hylobates syndactylus*
Features:	Uniform black fur; tailless
Habitat:	Tropical forests
Distribution:	Southeast Asia
Length:	Up to 90cm (35in)
Weight:	Up to 15kg (33lb)
Breeding:	One or two offspring after a 230- to 235-day gestation

PRIMATES: INDRIIDAE

Indri

The indri, also known on its native Madagascar as the babakoto, has a body length of up to 90cm (35.4in) and a weight of up to 10kg (22lb), making it the largest of the lemur species. Colouration is variable according to the individual, but a striking mix of jet black and pure white is common (often the rump, the neck and the underside of the limbs are white, the shoulders, hands, face and upper limbs being black). The face is surmounted by two black tufted ears. The indri's rear legs are very powerful for making long tree-to-tree leaps, and the hands have opposable thumbs, while the feet have opposable big toes. Indris spend most of their time in the trees, and when they do move across land they hold their arms up high and move using a bounding action. The indri social group is a male/female pairing and offspring, and in total numbers around two to five creatures. This grouping is matriarchal; the female is dominant and will eat before the male.

Species name:	*Indri indri*
Features:	Contrasting colour patterns on fur; black face; tufted ears
Habitat:	Rainforest
Distribution:	Eastern Madagascar
Length:	Up to 90cm (35.4in)
Weight:	Up to 10kg (22lb)
Breeding:	One or two offspring after a 120- to 150-day gestation

PRIMATES: INDRIIDAE

Verreaux's Sifaka

Verreaux's sifaka is another of Madagascar's beleaguered lemur population. It is classified as 'critically endangered' by the IUCN, the main problem lying in habitat destruction by humans for wood fuel and timber, and during crude slash-and-burn farming. The sifaka has an overall coat of white or grey fur, with a black face and dark crown. It can make leaps of astonishing distance between trees, generating thrust by its exceptionally long and powerful rear legs (some scientists think that a membrane stretched between the body and the upper arm also gives gliding assistance to the jump). Sifakas have mixed male/female/offspring groups that number up to 15 animals. During the breeding season (January to March), males will compete for access to the group's females, with violent fights occurring. The typical group home range is 1–9 hectares (2.5–22 acres), and the group will move around the whole territory feeding during a two- to three-week period. The usual diet is fruit, leaves, flowers, bark and seeds.

Species name:	*Propithecus verreauxi*
Features:	White or grey fur; dark face and crown; long limbs
Habitat:	Deciduous and evergreen forests
Distribution:	Madagascar
Length:	Up to 45cm (18in)
Weight:	Up to 5kg (11lb)
Breeding:	One or two offspring after a 131- to 160-day gestation

PRIMATES: LEMURIDAE

Mongoose Lemur

Mongoose lemurs are one of the smaller members of the lemur family, and are found on Madagascar and the nearby Comoro Islands. Their habits are varied, with the lemurs on the Comoro islands living in humid rainforest conditions, while those of Madagascar gravitate towards drier deciduous forests. Both males and females have grey bodies, although the female's coat is darker grey; and the males have red fur on the flanks and the face, while on the female these locations are white. The lemur social group is very small – usually just the male, the female and one or two offspring. The male will give his time to grooming and playing with the young, in addition to the care of the female. However, all mature offspring will be pushed out of the group to fend for themselves, usually when they are around two to three years old. Diet consists of leaves, flowers, fruit and pollen, and they will also take some agricultural crops. For this latter offence, they have been prejudicially hunted. Combined with habitat destruction, this hunting has reduced the population to danger levels.

Species name:	*Elemur mongoz*
Features:	Grey fur with red (male) or white (female) colouration on flank and face
Habitat:	Deciduous forests and rainforests
Distribution:	Madagascar and Comoro Islands
Length:	Up to 40cm (16in)
Weight:	Up to 3kg (6lb 10 oz)
Breeding:	One or two offspring after a 128-day gestation

PRIMATES: LEMURIDAE

Ring-tailed Lemur

The ring-tailed lemur is the most recognizable of the lemur species, known for the long catlike tail with its black-and-white rings. The back, flanks and outer limbs are covered with brown to fawn fur, the underparts and inner limbs are white, and the head features a grey cap and grey eye patches on a white base. Ring-tailed lemurs are much more comfortable on the ground than many other lemur species. When not foraging for food – the usual diet is fruit, leaves, bark, flowers and sap – they are often seen in an open sunbathing posture, spreading their limbs outwards to expose the chest and belly to the sun's heat. Ring-tailed lemurs form social groups of three–25 animals, and their societies are matriarchal. Not only does a female rule over the group, with other subordinate females arranged into a hierarchy, but also adolescent males are pushed out of the group while adolescent females are allowed to stay within their birth group.

Species name:	*Lemur catta*
Features:	Brown to fawn fur; black-and-white ringed tail; grey crown and eye patches
Habitat:	Forests and bush environments
Distribution:	Madagascar
Length:	Up to 46cm (18in)
Weight:	Up to 3.5kg (7lb 11oz)
Breeding:	One or two offspring after a 134- to 138-day gestation

PRIMATES: LEMURIDAE

Ruffed Lemur

The ruffed lemur's astonishing beauty and gregarious character have been its curse in recent years, with many being trapped and sold as pets, in addition to the hundreds killed for food. Consequently, as with many of Madagascar's lemurs, it sits on the IUCN's list of endangered species. Ruffed lemurs have black heads, feet and tails, with other black patches on the chest and flanks. The rest of the body is a luxurious white or red fur, although there are distinct variations among individuals and subspecies. In contrast to relatives such as the ring-tailed lemur, the ruffed lemur is almost exclusively arboreal, and rarely comes down from the trees. They are found in the wet evergreen forests concentrated in eastern Madagascar. From available evidence, the ruffed lemurs live in the typical social group of male and female plus offspring, and they are territorial – the females perform most the territorial defence. Fruit forms the bulk of the ruffed lemur's diet.

Species name:	*Varecia variegata*
Features:	White or reddish-white fur; black head, feet, tail flanks
Habitat:	Evergreen forests
Distribution:	Eastern Madagascar
Length:	Up to 55cm (22in)
Weight:	Up to 4.5kg (10lb)
Breeding:	One to three offspring after a 90- to 102-day gestation

PRIMATES: LORISIDAE

Lesser Bushbaby

Lesser bushbabies are tiny creatures that measure on average only 37cm (14.6in), including the tail, and weighing only 350g (13oz). They have grey or grey-brown coats, with whitish underparts and large, circular deep-hazel eyes. Their ears are large and flexible, the acute sense of hearing giving advanced alert of predators and prey. Found in the woodlands and bushlands of Central Africa, they tend to spend most of their time in the trees, and despite their small size they are known for their prodigious leaps between branches. Bushbabies have a broad diet. Acacia-tree gum provides a major source of nutrition, but the animals will also eat insects, small mammals, lizards and even scorpions. Bushbabies make nests by either building a nest in the treetops or using an abandoned bird's nest or animal burrow. They are territorial creatures, and the territories are marked out with urine – the bushbaby will 'wash' its feet in its own urine, then walk it around to establish scent boundaries.

Species name:	*Galago senegalensis*
Features:	Grey or grey-brown coat; white underparts; large eyes and ears
Habitat:	Woodlands and bushlands
Distribution:	Central Africa
Length:	Up to 37cm (14.6in) including tail
Weight:	Up to 350g (13oz)
Breeding:	Two offspring after a 125-day gestation

PRIMATES: PONGIDAE

Lowland Gorilla

The mightiest of all the primates, the gorilla is also one of the most endangered. There are three main subspecies: the western lowland gorilla (*Gorilla gorilla gorilla*), the eastern lowland gorilla (*Gorilla beringei graueri*) and the mountain gorilla (*Gorilla beringei beringei*), although some authorities now reclassify the subspecies as individual species under the family *Hominidae*. Gorillas are immensely powerful creatures, with huge limbs, strong and extremely flexible shoulders, opposable big toes, a short muzzle and a solid-looking forehead. They are covered in dense black hair (which is longer in mountain gorillas), and in mature males the hair on the back and rump turns silver-grey, which is the reason they are known as silverbacks (males are around twice the weight of females). Their habitat is tropical and montane forests, and their diet is leaves, fruit, insects, bamboo, roots and other plant foods, plus insects such as termites. Intimate social groups of 3–40 gorillas are formed, controlled by a dominant silverback, with a home range of 800–1800 hectares (2000–4450 acres). Communication within the group is highly advanced, with 'talking' by vocalizations, facial expressions and, when threatened, chest-beating and barking.

Species name:	*Gorilla gorilla*
Features:	Black coat; thick, powerful limbs; flat face with strong jawline and forehead
Habitat:	Tropical and montane forests
Distribution:	Isolated pockets of Central and eastern Africa
Length:	Up to 1.9m (6ft 2in)
Weight:	Up to 200kg (430lb)
Breeding:	One infant after an 8-month gestation

PRIMATES: PONGIDAE

Bonobo

Bonobos are also called pygmy chimpanzees, but the dimensions of the two are fairly similar apart from the bonobo being slightly shorter and slimmer in build; only the hind limbs are longer. They are concentrated in the Congo Basin, and form mixed social groups of males, females and offspring, the whole group numbering from three–10 animals (groups of up to 30 have been observed when food has been made artificially plentiful). A male will dominate the group, although a mature female will also be a potent presence. The males grow up within the same group from birth, while young adult females must leave to find their own group. Although the bonobo has much in common with the chimpanzee, there are some slight behavioural differences. The bonobo is a knuckle walker like the chimpanzee, but spends more time walking bipedally. Bonobos are very sexually active creatures, with sexual activity crossing into same-sex contexts, as well as between male and female pairings. This activity is probably a way of establishing social rank and harmony.

Species name:	*Pan paniscus*
Features:	Black coat; mostly black face; opposable thumbs; long limbs
Habitat:	Tropical forests
Distribution:	Central Africa
Length:	Up to 83cm (2ft 7in)
Weight:	Up to 39kg (86lb)
Breeding:	Usually one infant after a 240-day gestation

PRIMATES: PONGIDAE

Chimpanzee

Chimpanzees are found in forest habitats across western and Central Africa. They are renowned for their intelligence, using a range of more than 30 different vocalizations to communicate within the group. The 'pant-hoot' call – a loud shriek that transfers a range of information, from danger to the presence of food – can project over distances of 2km (1.2 miles). Chimpanzees are problem-solving creatures, using twigs and rocks to achieve various tasks, such as removing insects from their nests, and even using mashed-up leaves as sponges to soak up water. Their hairless faces are, like humans, ideally suited to nonverbal communication. The chimpanzee's day is spent mostly acquiring food. Chimpanzees are omnivorous – they eat fruit, flowers, leaves, seeds and some insects, but will also hunt down and kill other monkeys and even small ungulates, working as a team and literally tearing the prey apart. When not feeding, the chimpanzee group members (a community numbers from 20–100 animals) groom one another, then sleep the night in tree nests, each adult typically constructing a new nest every night.

Species name:	*Pan troglodytes*
Features:	Thin black coat; exposed flesh on face, ears, hands and feet; arms longer than legs; opposable toes and thumbs
Habitat:	Tropical forests
Distribution:	West and Central Africa
Length:	Up to 90cm (2ft 11in)
Weight:	Up to 60kg (130lb)
Breeding:	Usually one infant after an 8-month gestation

PRIMATES: PONGIDAE

Orang-Utan

Unlike many primate species, the orang-utan is mainly a solitary animal, although male and female territories may frequently overlap, and the animals will sometimes associate with one another during feeding times. It is almost entirely arboreal – males and females will even copulate while hanging from branches, and females give birth up in the tree canopy, and the infant then clings to the mother's fur. The body is superbly adapted for the habitat – the arms have a span of more than 2m (6ft 6in), while both hands and feet have opposable structures for a vicelike grip on branches. Orang-utans have a very shaggy red-brown coat, but only the males have the defining grey cheek pads that frame the face (these keep growing outwards as the animal ages). The future for the orang-utan is precarious. Only an estimated 20,000 survive in the tropical forests of Borneo and Sumatra, and their habitats are rapidly shrinking under the incessant depredations of logging and slash-and-burn farming.

Species name:	*Pongo pygmaeus*
Features:	Brown to deep-orange coat; long arms in relation to body; males: large cheek pads
Habitat:	Tropical forests
Distribution:	Borneo and Sumatra
Length:	Up to 1.4m (4ft 7in)
Weight:	Up to 80kg (175lb)
Breeding:	One or two infants after a 233- to 263-day gestation

PRIMATES: TARSIIDAE

Tarsier

The tarsier is a diminutive primate found in the tropical rainforests and bamboo forests of Southeast Asia, concentrated mainly in the Philippine Islands and in Indonesia. It grows to only 15cm (6in) in the body, and its main features are a small, round furry body, enormous eyes and bare toes and hands with gripping pads and claws. It is almost entirely arboreal in behaviour (it can hop only when on the ground), and is also nocturnal. It leaps from branch to branch in the almost total darkness of the rainforest, its large eyes providing excellent night vision, while its slender fingers and toes (there are five toes on each foot) give superb grip on slender branches. Tarsiers are found in several species, all carnivorous. They diet principally on insects, but will also take lizards and some small mammals if the need or opportunity arises. When hunting, it can turn its head 180 degrees to its chest, and it stalks and seizes the prey.

Species name:	*Tarsius spp.*
Features:	Typically brown fur with lighter belly; large eyes; pointed snout
Habitat:	Tropical forests
Distribution:	Southeast Asia
Length:	Up to 15cm (6in) without tail
Weight:	Up to 135g (4oz)
Breeding:	One infant after a 180-day gestation

PROBOSCIDEA: ELEPHANTIDAE

Asian Elephant

The Asian elephant is usually distinguished from its African relative by the much smaller ears and shorter male tusks (females frequently do not have tusks). The back is also flatter, this making the head the highest point of the body (on the African elephant it is the shoulders). The trunk shape also has a subtle difference at the tip: African elephants have a process (a pointed swelling, used for gripping) at both the top and the bottom of the tip, whereas the Asian elephant has only one process, at the top. A typical Asian elephant will grow up to 3.5m (11ft) in length, and weigh up to 5 tons (5.5 tonnes). Male and female Asian elephants have different social groupings. The females will gather together in herds of 15–30 animals, including offspring, while the males will form loose bachelor groups or travel alone, neither sex seeming to establish a territory. Within female groups, as with the African elephant, the calves remain very close to the mother and enjoy committed group protection from predators.

Species name:	*Elephas maximus*
Features:	Grey-brown body; long prehensile trunk
Habitat:	Tropical scrub forest
Distribution:	South and Southeast Asia
Length:	Up to 3.5m (11ft) without tail
Weight:	Up to 5 tons (5.5 tonnes)
Breeding:	One infant after an 18- to 22-month gestation

PROBOSCIDEA: ELEPHANTIDAE

African Elephant

The African elephant is the world's largest land mammal (and the largest elephant species), and is found throughout many parts of sub-Saharan Africa. The back slopes down from its huge head, and the head carries tusks that grow up to 2.5m (8ft) long and can each weigh 130kg (300lb). Each ear can be 1.2m (4ft) across, these being wafted in the air to accelerate heat loss. Herds are made up of females and offspring, with the males either forming their own small groups or entering the female herds for mating. Herd size can be very large – several hundred in number – and each of the adults will need to eat around 160kg (350lb) of leaves, grass, fruit, branches and bark every day, often resulting in major environmental destruction. The nimble trunk is used to grip food and perform other tasks, and it is so sensitive and manipulable at the tip that it can pick up individual leaves or pieces of grass. The trunk is also used to throw up dust and squirt water during bathing.

Species name:	*Loxodonta africana*
Features:	Grey-brown body; long prehensile trunk; large curving tusks
Habitat:	Scrub forest, forests, grasslands and arid plains
Distribution:	Sub-Saharan Africa
Length:	Up to 5m (16ft) without tail
Weight:	Up to 7 tons (7.7 tonnes)
Breeding:	One infant (sometimes two) after a 22-month gestation

RODENTIA: BATHYERGIDAE

Naked Mole-Rat

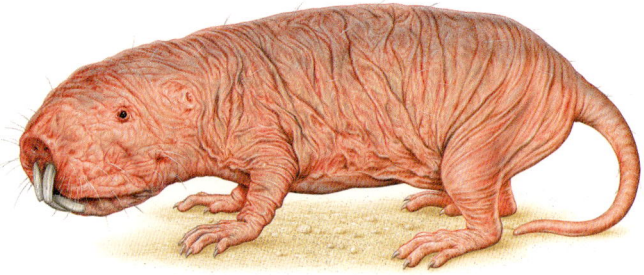

The naked mole-rat appears to be almost completely hairless to the naked eye, although it does have a very thin covering of pale hairs on its body. The whiskers and hairs on the tail help the mole-rat feel its way in darkness, while hairs between the toes assist in pushing aside loose earth. Mole-rats live almost entirely underground, working together in chains to dig tunnels and chambers, using strong limbs and extremely large incisor teeth (a quarter of the mole-rat's muscle power is concentrated in its jaw). The mole-rats form large colonies of nearly 100 animals, and the total length of the entire tunnel system can add up to several kilometres, the typical structure being a central chamber, food chambers and connecting tunnels. Socially the mole rat is unusual: the colony has only one dominant and breeding female, the rest of the colony being employed in tunnel digging and food gathering (the mole-rat diet consists of roots, tubers and bulbs). Despite this arrangement, the 'queen' can produce 20 infants per litter, and new litters about every 80 days.

Species name:	*Heterocephalus glaber*
Features:	Pink, fleshy body; minute eyes and ears; short, rounded tail
Habitat:	Beneath arid plains
Distribution:	East Africa
Length:	Up to 9cm (3.5in) without tail
Weight:	Up to 80g (3oz)
Breeding:	Up to 20 infants after a 70-day gestation

RODENTIA: CAPROMYIDAE

Cuban Hutia

The Cuban hutia, also known as Demarest's hutia, is a Cuban inhabitant that appears somewhere between a rat and a vole. It grows to a body length of up to 60cm (23.6in), with a tail measuring 15–26cm (6–10in). Its body is divided between dark upper parts (shades of brown, black and red) and paler underparts. Hutias inhabit most of the natural environments found on the island of Cuba, including montane and lowland forests, arid scrub, swamplands and coastal areas. Their habitat versatility is partly explained by a broad diet, which ranges from leaves and fruit to lizards. Cuban hutias are expert climbers, using their sharp claws for grip on branches and tree trunks. On the ground they walk with a waddling motion, although they can hop at speed when threatened by a predator. Both male and female hutias are territorial animals, marking out their home range using urine. Hutias come together in small groups for grooming and playing, although these groups number only around two or three animals.

Species name:	*Capromys pilorides*
Features:	Long tail; curved snout; dark fur on upper body, pale on underparts
Habitat:	Forests, scrub, swamplands, coastal semi-desert
Distribution:	Cuba
Length:	Up to 60cm (23.6in) without tail
Weight:	Up to 7kg (15lb)
Breeding:	One to three infants after a 120- to 126-day gestation

RODENTIA: CAPROMYIDAE

Coypu

The South American coypu is a semi-aquatic animal, inhabiting the shores and banks of lakes, rivers, streams and marshland. Within this terrain the coypus principally eat water vegetation, and have been seen sitting on floating logs used as feeding platforms, although more typically they make platforms of vegetation on which to feed and groom. Coypus are expert swimmers, powered by webbed rear feet; they can maintain dives for up to 10 minutes. The dens are dug into banks to depths of 15m (49ft), with several nesting chambers. A den will be inhabited by a group of 3–15 coypus, mostly made up of females and their offspring, with one dominant male. Females breed throughout the year, and produce on average two litters of infants, with each litter numbering up to 13 animals, although the average is more in the region of three to six. Coypus have been intensively farmed for their thick brown fur, and escaped coypus have established wild populations in North America, Europe and Asia.

Species name:	*Myocastor coypus*
Features:	Thick brown fur; eyes and ears set high on head; webbed rear feet
Habitat:	Aquatic environments
Distribution:	South America
Length:	Up to 58cm (23in) without tail
Weight:	Up to 6.5kg (14lb)
Breeding:	Two–13 infants after a 127- to 139-day gestation

RODENTIA: CASTORIDAE

North American Beaver

Beavers are the largest of the North American rodents, inhabiting territories ranging from Alaska down through to the borders of Florida. They live in aquatic environments, particularly alongside lakes, rivers and streams where there is plenty of available vegetation nearby. Beavers have thick waterproof coats (oiled with glandular secretions), usually of red-brown or black-brown colouration, and the tail is paddle-like, flat and scaly, and used to provide thrust when swimming. The head is very large, and contains the long upper incisors, which grow to around 2.5cm (1in) long. These are used for cutting through branches (the diet is aquatic bark plants) and for gnawing through the trunks of small trees. The beaver's lodge consists of a mass of intertwined sticks bound with mud, accessed by underwater entrances. Beavers also construct dams to slow water flow around the lodge. The beaver's social unit is a mixed male/female family of related animals, and the adult beavers will mark out their territory using mud piles and anal secretions.

Species name:	*Castor canadensis*
Features:	Reddish-brown to black fur; flat, paddle tail; large upper incisors
Habitat:	Inland aquatic environments
Distribution:	North America
Length:	Up to 88cm (35in) without tail
Weight:	Up to 26kg (57lb)
Breeding:	One to four kits after a three-month gestation

RODENTIA: CASTORIDAE

European Beaver

Populations of European beavers are scattered throughout continental Europe from Scandinavia down to the Balkans. They have much in common with the North American beaver, but their dimensions are larger – they can grow up to 1m (39in) in the body alone, with the tail adding another 30–38cm (12–15in). European beavers have similar social structures to their North American relative, living in family groups of around five to eight animals, consisting of monogamous parents (the female is dominant in beaver society) and one or two years' worth of offspring. Their lodges, however, vary in construction, depending on the local environment. While the European beavers do build stick and mud lodges, they will also dig lodges directly into the riverbank, protecting the entrances with 'castles' of sticks. Damming regulates the water levels around the lodges, and both North American and European beavers will create dams of considerable size, some approaching 32m (100ft) in length. The young will stay in the lodge for around the first two to three weeks of life before venturing outside.

Species name:	*Castor fiber*
Features:	Yellow-brown to black fur; flat, paddle tail; large upper incisors
Habitat:	Inland aquatic environments
Distribution:	Europe
Length:	Up to 1m (3ft 3in) without tail
Weight:	Up to 35kg (77lb)
Breeding:	One to four kits after a 105-day gestation

RODENTIA: CAVIIDAE

Mara

The mara, also known as the Patagonian cavy, lives in the grasslands and bushlands of southern South America. Although included in the cavy group of rodents, its lower-body appearance is more deerlike. It walks on long, slender legs, and it has two large ears set atop the head, which has more of a classic cavy shape. Its overall coloration is grey-brown, with white patches of fur on the throat and rump. The long legs provide various styles of motion, from a standard walk to a rabbit's hop, and maras can run at speeds of up to 45km/h (30mph). Male and female maras pair together in lifetime monogamous relationships in which the male is dominant, although he usually follows behind the female during foraging trips. Numerous pairs occasionally meet up around communal feeding sites (they are not territorial animals). Maras also create communal dens for the birth of young and for sleeping, and these can house up to 15 pairs of animals.

Species name:	*Dolichotis patagonum*
Features:	Grey-brown fur; white patches on throat and rump; very short tail; large ears
Habitat:	Grasslands and bushlands
Distribution:	Southern South America
Length:	Up to 78cm (31in) without tail
Weight:	Up to 2kg (4lb 8oz)
Breeding:	One to three young after a three-month gestation

RODENTIA: CHINCHILLIDAE

Chinchilla

Chinchillas have become extremely popular as domestic pets in recent decades. The wild population, however, is classified as 'vulnerable' by the IUCN, the depletion due mainly to detrimental crop production methods in some of its habitats. The chinchilla is a small, rounded rodent with thick silver-grey fur, cream underparts, strong rear legs, a very bushy tail and large rounded ears. The fur is essential protection in the chinchilla's chosen habitat – mountainous areas at elevations of 3000–5000m (9842–16,404ft), where the winter produces blizzards and subzero temperatures. Chinchillas diet mainly upon grasses and leaves, and they live together in colonies of up to 100 animals, using caves and similar features for shelter. Within the colony, females are the dominant sex, and are known for their aggression towards both males and other females during the breeding season. Chinchillas are also intensive breeders within their monogamous relationships, producing up to six offspring each year, with the young chinchillas becoming sexually active at about eight months old.

Species name:	*Chinchilla lanigera*
Features:	Silver-grey fur; cream underparts; bushy tail; large ears
Habitat:	Montane scrub
Distribution:	Southwest South America
Length:	Up to 23cm (9in) without tail
Weight:	Up to 500g (18oz)
Breeding:	Two or three young after a 111-day gestation

RODENTIA: DASYPROCTIDAE

Azara's Aguoti

The family *Dasyproctidae* contains 13 species of cavy-like rodents. A typical example is the Azara's aguoti (*Dasyprocta azarae*) of South America, which is found in territory stretching from Mexico to northern Argentina. Aguotis grow to around 50cm (20in) long and have sleek, speckled orange-brown coats, rounded muzzles and short ears. The legs are short, but provide good traction through three clawed toes on the hind feet, with five on the front. Aguotis are very nimble, able to sprint at speed or leap high from a standing start should danger approach (birds of prey and jaguars are the main predators). As with many creatures of its type, aguotis form monogamous male/female relationships, the initial active courtship occurring when the male sprays urine over the female, inducing her to make a courtship dance. The pair of aguoti will establish a territory of some 1–2 hectares (2.5–5 acres), and will spend their time together sleeping, food gathering and grooming each other's fur for parasites. Male aguotis will protect territorial integrity aggressively, engaging in violent fights with intruders.

Species name:	*Dasyprocta azarae*
Features:	Short orange-brown fur; short tail, ears and legs
Habitat:	Forest, brushland, grasslands, agricultural areas
Distribution:	South America
Length:	Up to 50cm (20in) without tail
Weight:	Up to 3kg (6lb 10oz)
Breeding:	Two to four young after a 104- to 120-day gestation

RODENTIA: DIPODIDAE

Desert Jerboa

The tiny desert jerboa is a common rodent from North Africa to western Asia. It is essentially a mouse on kangaroo-like rear legs – it moves by hopping, rather than walking. The hind legs – which can measure up to 7.5cm (3in) in an animal that grows to only 12cm (5in) in the head and body – also serve to get the jerboa out of trouble with spectacular leaps, and help to elevate the male jerboa's stature when trying to attract a mate. Jerboas are nocturnal animals, the darkness providing protection from both predators and the equatorial daytime heat. They live alone, and dig spiral burrows 1.2m (4ft) deep into the ground. The jerboa diet is mainly herbivorous, although insects will occasionally be eaten. Jerboas can travel long distances in search of food. These journeys – up to 10km (6 miles) – are dangerous because the jerboas attract the attention of foxes, snakes and vipers, although their speed is a good defence.

Species name:	*Jaculus jaculus*
Features:	Mottled grey to orange fur; white underparts; long hind legs and long tail
Habitat:	Arid regions
Distribution:	North Africa into western Asia
Length:	Up to 12cm (5in) without tail
Weight:	Up to 75g (2.6oz)
Breeding:	One to six offspring after a 25-day gestation

RODENTIA: ERETHIZONTIDAE

North American Porcupine

In North America, only the beaver is a larger rodent than the North American porcupine. The porcupine's body is covered in spiny quills and fur, with a concentration of long yellow-white quills extending out from the head and down the back, with brown-black fur along the flanks. The quills can grow up to 8cm (3in) long, and they are hollow in structure, with tiny barbs at the tips; they present a powerful disincentive to attacks by predators. Despite its size, the porcupine is a reasonable tree climber, the powerful claws for digging into bark and branches. While feeding in a tree – it will collect most edible plant structures – the porcupine is often seen gripping the tree with its hind legs while manipulating the food with its front hands, the quills acting as an additional anchor. The mating season can be a violent time for male porcupines, as they defend their breeding rights to a single female over several days, fighting off competitors in dangerous quill battles.

Species name:	*Erethizon dorsatum*
Features:	Dense covering of yellow to white quills and brown-black fur; clawed feet
Habitat:	Deciduous forests, tundra, grasslands and arid regions
Distribution:	North America
Length:	Up to 80cm (32in) without tail
Weight:	Up to 7kg (15lb)
Breeding:	One offspring after a 210-day gestation

RODENTIA: GEOMYIDAE

Pocket Gopher

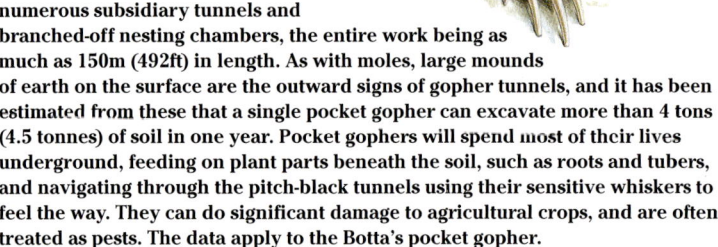

There are 35 different species of pocket gopher in the United States and South America. They are vole-like burrowing creatures that range in colour from light brown to black, and have short bare tails and large cheek pouches (hence the 'pocket' name), which are used to carry food. The pocket gopher's burrows are substantial feats of engineering for the animal; they consist of a central tunnel with numerous subsidiary tunnels and branched-off nesting chambers, the entire work being as much as 150m (492ft) in length. As with moles, large mounds of earth on the surface are the outward signs of gopher tunnels, and it has been estimated from these that a single pocket gopher can excavate more than 4 tons (4.5 tonnes) of soil in one year. Pocket gophers will spend most of their lives underground, feeding on plant parts beneath the soil, such as roots and tubers, and navigating through the pitch-black tunnels using their sensitive whiskers to feel the way. They can do significant damage to agricultural crops, and are often treated as pests. The data apply to the Botta's pocket gopher.

Species name:	*Thomomys bottae*
Features:	Grey-brown to orange fur; small ears (can be closed to dirt); flat head; long whiskers
Habitat:	Underground in forests, woodlands, deserts and grasslands
Distribution:	USA
Length:	Up to 30cm (12in) without tail
Weight:	Up to 55g (2oz)
Breeding:	Three to seven offspring after a 19-day gestation

RODENTIA: GLIRIDAE

Edible Dormouse

The edible dormouse is also known as the fat dormouse, and it takes the former name from being part of the autumnal diet of ancient Romans. In many ways the mouse more resembles a squirrel. Its back legs are very long, and enable the mouse to make dramatic leaps amid tree branches – distances of up to 7m (23ft) have been recorded. It can climb trees with ease, thanks to hard pads on all the clawed feet. The tail is also long and bushy, and is used for balancing. Edible dormice are primarily nocturnal creatures, and all of their senses are highly sensitive to aid night-time movement. As autumn draws in, edible dormice begin preparations for a seven-month hibernation, having already dug extensive underground tunnel networks during the summer. They eat intensively to store up fat, and during the hibernation the body temperature drops dramatically and respiration falls to around three cycles per minute. The dormouse will instantly awake, however, if a threat interferes with its hibernation chamber.

Species name:	*Glis glis*
Features:	Brown to grey fur; white underparts; dark eye patches; bushy tail
Habitat:	Forest environments
Distribution:	Central and southern Europe; western Asia
Length:	Up to 20cm (8in) without tail
Weight:	Up to 150g (5oz)
Breeding:	One to 11 offspring after a 30- to 32-day gestation

RODENTIA: HYDROCHAERIDAE

Capybara

Capybaras are enormous members of the rodent family, growing up to 1.3m (4ft 6in) in length. Despite being killed in large numbers for their meat and hide (on a commercial basis as well as through casual hunting), capybaras still remain common throughout much of South America. Capybaras have deep, heavy bodies atop very short but strong legs. The snout is long, with the narrow eyes set far back on the head, and the muzzle is blunt. Despite their size, capybaras are good swimmers – their hooflike feet have a webbed structure between the toes – and their habitats are typically areas of dense vegetation bordering on lakes, rivers and streams. Capybaras are diurnal animals, spending most of the day resting (usually by wallowing in shallow water) and feeding. They are entirely herbivorous, and eat grasses, water plants, bark and agricultural crops such as corn and melons. The social group consists of a dominant male, several females (also arranged in hierarchy) and offspring, the group numbering up to 20 animals.

Species name:	*Hydrochaerus hydrochaerus*
Features:	Brown to orange-brown fur; short legs; heavy muzzle
Habitat:	Along vegetation-heavy waterways
Distribution:	South America
Length:	Up to 1.3m (4ft 6in) without tail
Weight:	Up to 66kg (145lb)
Breeding:	One to eight offspring after a 150-day gestation

RODENTIA: HYSTRICIDAE

Cape Porcupine

The Cape porcupine has a formidable array of defensive black-and-white quills and spines on its back, which grow up to 50cm (20in) and can be made erect when threatened. Barbed tips on the quills and spines hook into the flesh of predators and detach from the porcupine's body (they grow back quickly), and some of the spines have a hollow rattling structure to provide an auditory deterrent. Combined with the animal's ability to make a backwards charge – or its defensive technique of lodging itself in a hole with only spines exposed above ground – it is not surprising that the Cape porcupine has few animal predators. Infant cape porcupines are born with their quills and spines, but these are soft until exposed to the air, after which they harden. Cape porcupines are primarily herbivorous, eating fruit and berries, or digging up roots and tubers, although they are also known to eat some carrion when necessary. Their activity is mostly confined to nocturnal hours, during which time they will wander for several miles hunting for food.

Species name:	*Hystrix africaeaustralis*
Features:	Black-and-white spines and quills on back and flanks, elsewhere black fur; short legs
Habitat:	Forests, woodlands, grasslands, montane brush
Distribution:	Central and southern Africa
Length:	Up to 80cm (32in) without tail
Weight:	Up to 20kg (44lb)
Breeding:	One to four offspring after a six- to eight-week gestation

RODENTIA: HYSTRICIDAE

North African Porcupine

The North African porcupine is found not only in North Africa, but also in parts of mainland Italy and Sicily. It has dense crests of quills and spines on its head, nape of the neck, back and rump, these being black-and-white mixed with black bristly hairs that cover the limbs, flanks and chest, with the exception of a white throat patch. With such a formidable armament, mating presents something of a challenge – the female must lift her tail high, while the male stands on his back legs and uses his front legs on the female's back to support his weight. Male and female pairings are monogamous, and the parents, with their offspring, will live together in an underground burrow (although the female will often give birth in a separate burrow to the family home). Many African porcupine species, including *Hystrix cristata*, are endangered animals because of hunting – their meat is a delicacy in many African communities. Big cats and hyenas will also attack porcupines, although even such large predators have been known to suffer serious injury or even death as a result of the porcupine's spines.

Species name:	*Hystrix cristata*
Features:	Dense coverings of spines and quills; black head, flanks and limbs; white throat patch
Habitat:	Forests, woodlands, grasslands, montane brush, deserts
Distribution:	North Africa, mainland Italy and Sicily
Length:	Up to 93cm (37in) without tail
Weight:	Up to 30kg (66lb)
Breeding:	One or two offspring after a 112-day gestation

RODENTIA: MURIDAE

Water Vole

The water vole is common across most of Europe and up into the northern reaches of eastern Russia and Siberia. As its name suggests, its preferred habitat is alongside inland waterways, and it is an excellent swimmer and diver, despite having none of the physical adaptations of an aquatic animal. However, water voles are also found in non-aquatic environments, such as meadows and woodlands, although these creatures tend to be smaller. Water voles dig extensive burrow networks into riverbanks. Attached to the labyrinthine tunnels – which can have a total length up to 70m (300ft) – are various nesting chambers and larders, the latter being particularly important for winter food storage. They tend to be crepuscular animals, feeding off plant foods within a territory of around 200 sq. m (2152 sq. ft). The territories are marked out using emissions from flank glands. Water voles have grey, brown and black colours in their coats, and like many rodents they have to gnaw hard materials to wear down their constantly growing front incisors.

Species name:	*Arvicola terrestris*
Features:	Short, coarse grey, brown and black fur; long tail; small ears
Habitat:	Alongside inland waterways; meadows and gardens
Distribution:	Europe, western Asia and northern Asia
Length:	Up to 23cm (9in) without tail
Weight:	Up to 300g (11oz)
Breeding:	Average four to six offspring after a 21-day gestation

RODENTIA: MURIDAE

Bank Vole

Bank voles have a wider range of habitat types than water voles; they are found in all types of European and North Asian forest, as well as scrublands, hedgerows and anywhere with dense vegetation. They are extremely common animals, and as such form an important part of the carnivorous food chain – weasels, stoats, mink, foxes and birds of prey all consume large numbers of bank voles. The bank vole evades predation by diving into thick undergrowth or into its underground burrows. Bank voles' fur is reddish brown on the back and forehead, lightening to grey along the flanks. The feet are white. They have a broad diet, including plant foods such as fungi and grass seeds through to insects and birds' eggs. They are active at any part of the day or night, although the twilight hours are preferred. Bank vole society is matriarchal, and females will usually establish a lifelong and aggressively defended territory, whereas males will leave their birth territory once they are mature.

Species name:	*Clethrionomys glareolus*
Features:	Red-brown fur lightening to grey on flanks and underparts; long tail with small bush at tip
Habitat:	Any vegetation-rich environment
Distribution:	Europe and northern Asia
Length:	Up to 13cm (5in) without tail
Weight:	Up to 35g (1.2oz)
Breeding:	One to 10 offspring after a 21-day gestation

RODENTIA: MURIDAE

Northern Pygmy Gerbil

There is a total of 110 species of gerbil distributed throughout Africa and Asia, with up to 62 of these classified as northern pygmy gerbils. Pygmy gerbils tend to be smaller than many of the other species of gerbils, but share most of the characteristics of the gerbil family. Gerbil habitats are diverse, although they prefer open land with patches of vegetation cover. The coat colour varies according to the species, but ranges from plain grey to orange-brown; the underparts are usually white. Depending on the species, gerbils can be intensive breeders, with some producing up to 13 offspring after a gestation lasting only 21–28 days. Social life is as diverse as the species—some gerbils living solitary lives, others in large colonies. All gerbils are burrowers, and will live in extensive underground tunnel networks and nesting chambers. They are quite vocal, using squeaks and chatterings, and, when alarmed, they will drum the hind feet on the floor to signal danger. The data below are for the Cheeseman's gerbil.

Species name:	*Gerbillus cheesmani*
Features:	Sand-coloured fur; long ears; tail longer than body length
Habitat:	Rocky terrain, scrubland
Distribution:	Middle East
Length:	Up to 13cm (5in) without tail
Weight:	Up to 63g (2oz)
Breeding:	Four to eight offspring after a 21-day gestation

RODENTIA: MURIDAE

Jird

Jirds are closely related to gerbils, but are distinguished from gerbils by belonging to the genus *Meriones*. There are 14 species of jird, which are found in a broad swathe of territory that includes Turkey, North Africa, Central Asia and southern Asia. A classic example of a jird, and one commonly owned as a pet, is known as the Mongolian gerbil (*Meriones unguiculatus*). Found in parts of China, Siberia and Mongolia, it is a hardy animal that survives the subzero winters and scorching summers of the region. Mortality among such creatures is high due to exposure and predation, but during the February to October breeding season a single female will typically produce around 28 offspring, split between three litters. Typical of all gerbils and jirds, the Mongolian gerbil creates extensive burrows and feeds off grasses and seeds. The Mongolian gerbil is also highly resistant to dehydration, concentrating its urine and faeces to reduce water loss. Mongolian jird communities number up to 20 animals and are led by a single dominant male.

Species name:	*Meriones unguiculatus*
Features:	Brown fur; light underparts; long whiskers; long hind legs
Habitat:	Rocky terrain, scrubland, grasslands and mountain valleys
Distribution:	East Asia
Length:	Up to 13cm (5in) without tail
Weight:	Up to 63g (2oz)
Breeding:	One to 12 offspring after a 19- to 30-day gestation

RODENTIA: MURIDAE

Harvest Mouse

The harvest mouse is the smallest rodent in Europe, its body length measuring a maximum of only 8cm (3in), with a tail of roughly the same length. Distribution, however, goes well beyond Europe, with harvest mice being found as far east as North Korea and southern China. It is an especially nimble mouse; the tail is prehensile, and this provides additional grip when climbing the thinnest blades of grass. Its preferred habitat is areas of dense vegetation growth, particularly hedgerows, deep grasslands, reedbeds, marshes and cornfields. It actually builds its own breeding nest, a spherical ball of woven grasses about 10cm (4in) in diameter suspended amid the vegetation at a height of 20–80cm (8–31in) from the floor. (During the winter, they nest underground.) Their nesting habits and their dependence on thick vegetation have meant that harvest mice populations have been affected by modern intensive agriculture, although they are now classified as 'lower risk' by the IUCN.

Species name:	*Micromys minutus*
Features:	Reddish-brown fur; white underparts; small ears; prehensile tail
Habitat:	Hedgerows, grasslands, reedbeds, marshlands and cornfields
Distribution:	Europe to East Asia
Length:	Up to 8cm (3in) without tail
Weight:	Up to 7g (0.2oz)
Breeding:	One to seven offspring after a 17- to 19-day gestation

RODENTIA: MURIDAE

Field Vole

The field vole (also called the short-tailed vole) has a mouselike appearance, although its ears and tail are shorter than most mouse species. Fur colour is yellow-brown, with pale underparts, and the preferred habitats are areas of dense vegetation that are damp – marshes, the borders of rivers, streams, lakes and ponds, wet meadows and moorland. They are highly territorial creatures, and the males – who establish exclusive territories using scent trails – will fight vigorously any male who intrudes into the home range. Female home ranges often overlap, and there is less antagonism. The centre of a home range will usually be a tussock of grass; this forms the nesting area and is served by numerous runways through the surrounding vegetation. The diet is principally grass and leaves, and feeding tends to be a night-time activity in the summer, moving to daytime during the winter months. Field voles are preyed on by many mammal species and birds of prey.

Species name:	*Microtus agrestis*
Features:	Yellow-brown fur; pale underparts; small eyes and ears
Habitat:	Hedgerows, grasslands, marshlands and borders of aquatic environments
Distribution:	Europe to East Asia
Length:	Up to 13cm (5in) without tail
Weight:	Up to 55g (2oz)
Breeding:	Four to six offspring after a 18- to 20-day gestation

RODENTIA: MURIDAE

House Mouse

The house mouse is one of nature's success stories. It is found everywhere in the world except the polar regions. Its population and distribution have expanded alongside those of humans, as it thrives among human habitations where it finds shelter and food (consequently it is also global pest). Indeed, house mice can struggle to survive in a wild environment, suffering from predation and from the competition over food with hardier rodents. It will usually, therefore, build tunnel nesting sites in those natural areas (fields, woodlands, hedgerows) that border human settlements, developing runways to and from the human dwelling for night-time feeding (they are nocturnal creatures). Its diet is totally omnivorous, and it will eat any digestible item it can hold. House mice are prolific breeders – a female is capable of producing more than 100 offspring every year in up to 10 litters. Only high mortality rates (a house mice typically survives less than a year in the wild) keeps the population from reaching crisis proportions.

Species name:	*Mus musculus*
Features:	Brown fur; large, rounded ears; long tail
Habitat:	Natural habitats bordering human habitations
Distribution:	Global except for polar regions
Length:	Up to 10cm (5in) without tail
Weight:	Up to 35g (1oz)
Breeding:	Three to 12 offspring after a 19- 21-day gestation

RODENTIA: MURIDAE

Muskrat

The muskrat originated in North America, but was introduced into Europe in 1905. Today it is found across much of the northern hemisphere, where it inhabits areas around lakes, rivers, streams and other inland waterways. In both appearance and behaviour, it is like a smaller version of the beaver. It has a dense coat of brown/red-brown fur and a flattened, scaly tail that provides a rudder when swimming. The feet are webbed, and the muskrat can shut its nose and ears when diving (dives are known to last for up to 17 minutes). Muskrats also have advanced body-temperature control, and can divert heat away from the extremities when underwater or in cold conditions. Muskrat territories are marked out with musk secretions, faeces and urine, and they live in large, often chaotic and violent, family groups. Like a beaver, they will construct lodges of sticks and mud, but also nest in tunnels bored directly into riverbanks. The diet is mainly aquatic plants, but they will eat fish and frogs.

Species name:	*Ondatra zibethicus*
Features:	Thick brown/red-brown fur; flattened tail; webbed feet; small, rounded ears
Habitat:	Inland aquatic environments
Distribution:	North America, Europe and Asia
Length:	Up to 35cm (14in) without tail
Weight:	Up to 2kg (4lb 6oz)
Breeding:	One to three offspring after a 29- to 30-day gestation

RODENTIA: MURIDAE

Grasshopper Mouse

Grasshopper mice are almost entirely carnivorous, and are efficient and aggressive hunters. There are three principal species of grasshopper mouse found in the United States and Mexico, and they tend to inhabit grasslands, hedgerows and fence borders. Typical prey includes beetles, spiders, grasshoppers and pill bugs, but they also kill scorpions, other mice, prairie voles and some rat species. The killing tool is the mouse's long incisor teeth, and they attack by ambush – they wait until the prey is within range, then pounce on it, grasp it with their sharp claws and kill it with a strong bite to the neck.

Plant food, such as seeds and grasses, is eaten, but most estimates place this at only 10 per cent of the animal's total diet. Socially the grasshopper mouse lives either alone or in male/female pairs, but cannibalism is common, and pairings will frequently end in a violent death for one of the creatures. Grasshopper mice are also known for a piercing shriek that is audible 100m (328ft) away, this being a signal of aggression. Data are for the southern grasshopper mouse.

Species name:	*Onychomys torridus*
Features:	Grey to cinnamon upper fur; white underparts
Habitat:	Prairie and scrub
Distribution:	USA and Mexico
Length:	Up to 13cm (5in) without tail
Weight:	Up to 125g (4oz)
Breeding:	One to six offspring after a 26- to 35-day gestation

RODENTIA: MURIDAE

Brown Rat

The brown rat ranks alongside the house mouse and black rat as being one of the world's most widely distributed mammals, after humans. They are large rodents, growing up to 28cm (11in) in the body, with a long, scaly tail adding another 23cm (9in). Their habitats were traditionally diverse, but today they are found in greatest concentrations around human habitations, where they feed and shelter opportunistically. However, they are also common to grasslands and alongside rivers, streams and lakes – they are excellent swimmers, the feet driving the animal across the water, while the tail is held aloft for balance. Brown rats are omnivorous, foraging for anything edible and attacking and killing rabbits, mice, birds, fish and lizards (larger prey is usually handled by a pack of rats). The brown rat's most refined senses are hearing and smell, but it is also acutely sensitive to ground vibrations. They are aggressively territorial, and will attack and often kill intruders (the pack identifies itself through scent signatures).

Species name:	*Rattus norvegicus*
Features:	Short brown to grey-brown coat; scaly, almost naked tail, bare feet
Habitat:	Urban, woodlands, aquatic environments, grasslands
Distribution:	Global except for polar regions
Length:	Up to 28cm (11in) without tail
Weight:	Up to 0.5kg (1lb 5oz)
Breeding:	Six to nine offspring after a 22- to 24-day gestation

RODENTIA: MURIDAE

Black Rat

Black rats have the same global distribution as brown rats, but generally are more common to tropical than temperate regions. They are also called ship rats on account of their liking for coastal regions and sailing vessels – transoceanic travel is a key reason for the spread of the black rat population. Black in colour, including the tail, and generally smaller than the brown rat, black rats live in large packs of up to 60 animals, led by dominant males – although females tend to show more aggression. Because of intense population densities, black rat home ranges are usually no more than 100 sq. m (1076 sq. ft), and the rats will violently protect the territory against intruders – the territory will contain nesting sites and a primary food source. They are nocturnal and mainly herbivorous, but will eat carrion, insects and refuse. Black rats are extremely nimble and are good swimmers. They are also spreaders of disease (black rats were responsible for the bubonic plague), so they are classified as a pest.

Species name:	*Rattus rattus*
Features:	Short black coat; scaly black tail, sometimes longer than the body
Habitat:	Urban, maritime vessels, grasslands
Distribution:	Global except for polar regions
Length:	Up to 24cm (9.4in) without tail
Weight:	Up to 250g (9oz)
Breeding:	Six to 12 offspring after a 21- to 29-day gestation

RODENTIA: MYOXIDAE

Hazel Dormouse

The hazel dormouse is a European species that inhabits deciduous and coniferous forests and woodlands. The habitat types yield a diet of fruit and nuts (hazelnuts are a preferred food), with occasional insects, fledglings and birds' eggs providing protein supplements. Green plant foods are usually avoided, as the hazel dormouse does not have a caecum, the part of the intestines that is used for digesting cellulose. Hazel dormice are good climbers and spend much time in the trees, either collecting food or nesting in a tree hole or in thick vegetation. They hibernate in the winter for seven months (October to April), curling up in a nest bedded with grass, shredded leaves and moss, the vegetation bound together with saliva. The nest will also contain food supplies built up during the summer and autumn, although the dormouse will need good body fat reserves to survive the winter. Dormice are territorial and solitary, and a male's home range can be up to 1 hectare (2.5 acres) in size (female home ranges are slightly smaller).

Species name:	*Muscardinus avellanarius*
Features:	Upper coat of mixed colours, yellow, red-brown; large eyes; bushy tail
Habitat:	Forests and woodlands
Distribution:	Europe
Length:	Up to 8.5cm (3in) without tail
Weight:	Up to 35g (1.2oz)
Breeding:	Two to seven offspring after a 21- to 24-day gestation

RODENTIA: PEDETIDAE

Springhare

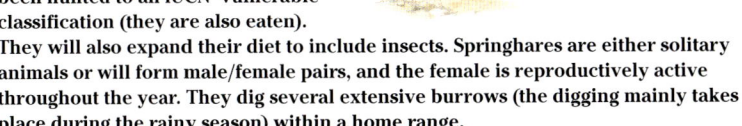

Generally described as a cross between a kangaroo and a rabbit, the springhare inhabits the southern third of Africa, its habitats being grasslands and arid scrublands. It has extremely large, clawed hind feet, which it uses to make a kangaroo-like leaping motion, with individual jumps of up to 2m (6ft 6in). The forelimbs are much less developed, and are principally used for holding food and for digging tunnels. The ears are large and the hearing acute, and the head is short with a blunt muzzle. Springhares have a fondness for agricultural crops, particularly wheat, barley and corn, and so they have been hunted to an IUCN 'vulnerable' classification (they are also eaten). They will also expand their diet to include insects. Springhares are either solitary animals or will form male/female pairs, and the female is reproductively active throughout the year. They dig several extensive burrows (the digging mainly takes place during the rainy season) within a home range.

Species name:	*Pedetes capensis*
Features:	Grey-brown upper fur; white underparts; large hind legs and ears; smaller forelimbs; black-tipped tail
Habitat:	Plains and arid regions
Distribution:	Central and southern Africa
Length:	Up to 40cm (16in) without tail
Weight:	Up to 4kg (8lb 4oz)
Breeding:	One offspring after a 78- to 82-day gestation

RODENTIA: SCIURIDAE

Black-tailed Prairie Dog

The black-tailed prairie dog is confined to a narrow corridor reaching from southwestern Canada to northern Mexico. The species has been heavily persecuted owing to its liking for cereal crops and its competition for food with grazing animals, but numbers appear to be stabilizing. They are, as their name suggests, plains dwellers, and they form into huge colonies known as 'townships', which number hundreds and even thousands of animals within an area of up to 65 hectares (160 acres). However, the smallest building-block of the townships is a male/female pair plus offspring, known collectively as a 'coterie'. A coterie will inhabit a burrow system that is defended territorially by both males and females, as it contains both food stores and infants. However, females will also fight within the coterie during the breeding season, when they have heightened aggression, and they are even known to kill the offspring of other females in the area. The subsequent birth of the young seems to quench the aggression.

Species name:	*Cynomys ludovicianus*
Features:	Red-brown upperparts; white underparts; black tail tip; small ears
Habitat:	Plains and grasslands
Distribution:	Western North America
Length:	Up to 30cm (12in) without tail
Weight:	Up to 1.5kg (3lb 5oz)
Breeding:	One to six offspring after a 33- to 38-day gestation

RODENTIA: SCIURIDAE

Southern Flying Squirrel

Southern flying squirrels have a patagium – a membrane of skin stretched between the wrist and the ankle, which acts as a glider wing when spread open. To make a glide – which the squirrel usually does to escape predators or to move to a different feeding location – the squirrel ascends up a tree as high as possible, then leaps off and spreads the patagium. It glides through the air for up to 80m (262ft), depending on launch height, and cushions the landing by lifting the tail to flatten the membrane into a more parachute-like angle. Southern flying squirrels are omnivorous creatures, eating insects, small mammals and birds, as well as nuts, acorns, bark, fruit, berries and seeds. They are nocturnal, the night being the best time to avoid the squirrel's principal predators – birds of prey, snakes, weasels, racoons and cats. Social structures range from male/female pairs through to larger groups of up to 20 animals. Dens and nests are made in tree hollows.

Species name:	*Glaucomys volans*
Features:	Patagium between wrists and ankles; grey fur on back and tail; white underparts; large eyes
Habitat:	Woodlands
Distribution:	Southeastern Canada down to Central America
Length:	Up to 15cm (6in) without tail
Weight:	Up to 85g (3oz)
Breeding:	One to six offspring after a 40-day gestation

RODENTIA: SCIURIDAE

Alpine Marmot

Alpine marmots are members of the squirrel family, but live on the ground rather than up in the trees. They have tan through red to grey-brown coats, with solid bodies and short, powerful legs suited to digging into hard, frozen earth (all the digits have claws apart from the thumb, which has a nail). Both tails and ears are short to minimize heat loss in the mountainous climate. Alpine marmots live in grassy and rocky terrain in various European mountain ranges, including the Alps, the Pyrenees, the Carpathians and the Tatras. Their diet consists of the thin high-altitude vegetation, typically grasses, leaves, flowers, bulbs and seeds, although they will also take insects. Alpine marmots live together in large family groups of up to 20 animals. This social system is very supportive: the marmots will spend much time grooming each other, and individual marmots will stand on guard duty while the others feed and play, emitting a piercing whistle if danger is spotted. During the winter, the marmots will hibernate in their underground burrow.

Species name:	*Marmota marmota*
Features:	Thick red to grey-brown fur; rounded face; short limbs and tail
Habitat:	Montane grasslands
Distribution:	Southern, central and eastern European mountain ranges
Length:	Up to 18cm (7in) without tail
Weight:	Up to 8kg (17lb 10oz)
Breeding:	Two to five offspring after a 33- to 34-day gestation

RODENTIA: SCIURIDAE

Woodchuck

The woodchuck, also known as the groundhog, is an extremely large member of the squirrel family, growing up to 52cm (20.5in) with a 11cm (4in) tail. It is distributed widely across North America, with its habitats being forests and woodlands, or open countryside that borders onto wooded areas. Woodchucks are burrowing animals, and in the autumn they use their clawed limbs and teeth (including their incisors to cut through roots) to dig a tunnel system featuring one to five entrances. The tunnel leads down to a main nesting chamber lined with grasses and leaves. Here they hibernate throughout the winter, occasionally waking on days of milder weather. Woodchucks wake from hibernation in February, stirred by warmer temperatures – hence 2 February is Groundhog Day in the United States, the day when the woodchuck is reputed to see if spring has arrived. The breeding season starts in the same month, and male woodchucks will fight violently for breeding rights.

Species name:	*Marmota monax*
Features:	Thick black-brown fur; white nose; bushy tail; short, strong limbs
Habitat:	Forests, woodlands and grasslands
Distribution:	North America
Length:	Up to 52cm (20.5in) without tail
Weight:	Up to 5kg (11lb)
Breeding:	One to nine offspring after a 31- to 32-day gestation

RODENTIA: SCIURIDAE

Eastern Grey Squirrel

The grey squirrel originally hailed from the United States and Canada, but through introduction measures in the nineteenth and twentieth century it is now a common rodent in the United Kingdom, where it has become a pest species. Grey squirrels are durable creatures that live in deciduous and mixed forest and woodland habitats. They have also adapted to living close to humans, and hence are frequent visitors to parks and gardens. Grey squirrels are voracious herbivores, and can do significant damage within their habitats. Their diet mainly consists of nuts, acorns, roots, shoots, seeds, flowers and many other plant foods, and they will strip bark to line their nests. Agricultural crops are also targeted in the winter months, lending further to its pest status. Grey squirrels are active the year round and are diurnal. They are territorial and can be aggressive towards each other; this is especially true of lactating females and males attempting to establish sexual dominance over other males.

Species name:	*Sciurus carolinensis*
Features:	Grey back, sides and limbs; long bushy tail; white underparts
Habitat:	Forests, woodlands, parkland and gardens
Distribution:	North America, United Kingdom and Italy
Length:	Up to 28cm (11in) without tail
Weight:	Up to 700g (25oz)
Breeding:	Two to eight offspring after a 42- to 45-day gestation

RODENTIA: SCIURIDAE

Eurasian Red Squirrel

The Eurasian red squirrel is found in a territorial band stretching from the United Kingdom to the eastern shores of China, although in many places its populations are endangered. The causes of its decline include the fact that it has been intensively hunted for fur, particularly in its Russian habitats, and also its competition with stronger species – the introduction of the large and aggressive grey squirrel into the United Kingdom in the early 1900s has almost destroyed British red squirrels. Red squirrels are separated from grey squirrels partly through a red-brown coat colour, although the fur may also be grey or grey-brown, particularly during the winter months. The prominent tufts on top of its ears are a changeless distinction, however. Like most tree squirrels, the red squirrel is an arboreal acrobat, its sharp claws providing grip, while the long bushy tail gives balance. Red squirrels are solitary apart from females with offspring. The young are born in a nest constructed from twigs in a tree fork, or in a vegetation-lined tree hole.

Species name:	*Spermophilus parryii*
Features:	Beige fur; white spots on side
Habitat:	Forests, woodlands, parkland and gardens
Distribution:	Europe to East Asia
Length:	Up to 25cm (10in) without tail
Weight:	Up to 475g (17oz)
Breeding:	Two to seven offspring after a 42- to 45-day gestation

RODENTIA: SCIURIDAE

Arctic Ground Squirrel

Arctic squirrels are the hardiest of creatures, inhabiting Arctic tundra habitats from northern Canada, Alaska and into Siberia. They survive the worst of the Arctic winter through hibernation, going into hibernation in September and returning to activity the following April, by which time they may have suffered a 30 per cent weight loss. They live precarious lives, falling prey to wolves, foxes, bears and birds of prey. To reduce their visibility on the barren tundra, they walk with their bodies pressed close to the ground. Diet consists of almost any plant material available, the ground squirrel using its sharp incisors to cut through branches and tough roots. Arctic ground squirrels form colonies of up to 50 animals, with dominant males controlling small territories within the colony, which are marked by scent glands. Births usually occur around June, and the pups are weaned in about six weeks. It is then imperative that they eat heavily to put on weight for the fast-approaching winter.

Species name:	*Spermiphilus parryii*
Features:	Red fur on face and sides; grey fur on back; pale underparts; long bushy tail; small ears
Habitat:	Arctic tundra
Distribution:	Far North America and Siberia
Length:	Up to 50cm (20in)
Weight:	Up to 800g (28oz)
Breeding:	Five to ten pups after a 25-day gestation

RODENTIA: SCIURIDAE

Eastern Chipmunk

The eastern chipmunk is a solitary animal, males and females associating only during one of the two breeding seasons during the year (February to April and June to August). These seasons lead to violent competition between the males for breeding rights, although ironically a female may choose to fight off a suitor even if he has won out over his rivals. Eastern chipmunks live in burrows within a small territory, and use these to store up large quantities of food for consumption during the winter months – they do not spent the entire winter in hibernation, but wake up intermittently to feed, and even venture outside. Chipmunks are a familiar site around parks and gardens, with their red-brown fur and five dark stripes running from the neck down the back. The tail is bushy, and there are white borders around the eyes. They have large pouched cheeks, which they stuff with seeds and other plant foods during their daytime foraging, and they are also known for their 'chip, chip, chip' alarm calls.

Species name:	*Tamias striatus*
Features:	Red-brown fur; five dark stripes running lengthwise; bushy tail
Habitat:	Woodlands and forests
Distribution:	Eastern North America
Length:	Up to 16.5cm (6.5in)
Weight:	Up to 125g (4oz)
Breeding:	One to nine offspring after a 31-day gestation

SIRENIA: DUGONGIDAE

Dugong

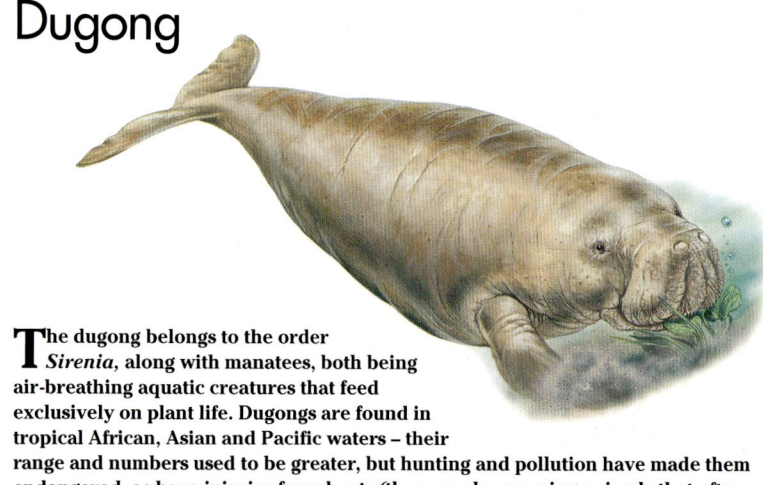

The dugong belongs to the order *Sirenia*, along with manatees, both being air-breathing aquatic creatures that feed exclusively on plant life. Dugongs are found in tropical African, Asian and Pacific waters – their range and numbers used to be greater, but hunting and pollution have made them endangered, as have injuries from boats (they are slow-moving animals that often float on the surface). Dugongs grow up to 4m (13ft) in length, with a large crescent tail, rounded foreflippers, a steeply angled snout and an upper lip that protrudes over the lower jaw. They have a single pair of tusklike incisors (dugongs have also been hunted for their ivory), and these are used to gather up sea grasses from the sea floor. Feeding is performed in correspondence with tidal patterns in shallow waters usually no more than 5m (16ft) deep, and the dugong often uses its foreflippers to walk across the bottom. Dugongs gather in groups that typically number 10–20, but which are occasionally much larger (several hundred).

Species name:	*Dugong dugon*
Features:	Grey-brown skin; crescent tail; paddle-like flippers; overhanging upper jaw
Habitat:	Rivers and coastal waters
Distribution:	Coastlines from eastern Africa to Pacific
Length:	Up to 4m (13ft)
Weight:	Up to 900kg (1985lb)
Breeding:	One calf after a 12-month gestation

SIRENIA: TRICHECHIDAE

West Indian Manatee

There are three species of manatee living in South American, US and African waters. The West Indian manatee inhabits coastal seas, rivers and estuaries ranging from Florida down to northeastern Brazil, and it is capable of changing from freshwater to saltwater habitats without problem. At their furthest, manatees have been found 200km (124 miles) inland. Manatees are huge creatures weighing up to 600kg (1320lb). Their skin is grey or grey-brown, and it is constantly shedding the surface layer, which helps to prevent excessive build-up of algae and parasites. Typical of sirenians, the West Indian manatee drifts slowly in shallow waters, feeding from plant life (and the occasional fish), which it pushes into its mouth with its flippers; it must return to the surface at least once every 20 minutes to breathe. Manatees form large non-hierarchical groups; females, when in season, may be pursued by up to 20 males competing for breeding rights.

Species name:	*Trichechus manatus*
Features:	Grey-brown skin; rounded jaw; thin hair covering
Habitat:	Rivers, estuaries and coastal waters
Distribution:	Southeastern USA to Brazil
Length:	Up to 4.5m (15ft)
Weight:	Up to 600kg (1320lb)
Breeding:	One calf after a 12- to 14-month gestation

TUBULIDENTATA: ORYCTEROPODIDAE

Aardvark

The aardvark – the name is the Afrikaans term for 'earth pig' – has a piglike body with an unusually long snout that terminates in a blunt muzzle, and a head topped by two enormous ears. It is a superb digger, creating burrows using its heavy, clawed front feet to cut away the earth, pushing the deposits backwards with the rear legs. While digging, the aardvark's ears are folded backwards to keep out dirt, while the nostrils are also protected by a thick filling of nostril hair. It makes burrows 3–10m (10–23ft) long, and at full pace an aardvark can burrow faster than several people with shovels. The burrows are set within a 2–5 sq. km (0.7–2 sq. mile) home range, across which the aardvark establishes a network of paths for its nocturnal feeding activities. Aardvarks feed on termites and ants, licking them up with a long tongue and swallowing them; the molar teeth are largely non-functional. Ants' nests and termite mounds are broken open with the claws before feeding.

Species name:	*Orycteropus afer*
Features:	Thin brown fur; long snout with flattened nostrils; large ears; powerful claws
Habitat:	Grasslands and arid regions
Distribution:	Sub-Saharan Africa
Length:	Up to 1.6m (5ft 3in) without tail
Weight:	Up to 64kg (142lb)
Breeding:	One offspring after a 243-day gestation

XENARTHA: BRADYPODIDAE

Maned Three-toed Sloth

The maned three-toed sloth is an arboreal creature that moves with great slowness through the tropical forests of eastern Brazil. Its limbs are powerfully developed, but overall muscle mass is poor and struggles to cope with the size of the sloth's body. The sloth also has trouble regulating its body temperature on account of a poor metabolism (it is 40 per cent slower than similar-sized mammals), so it will often ascend to the top of the tree canopy to expose itself to the sun to regulate its temperature. Its main survival skill is an ability to heal quickly when wounded; severe flesh wounds usually heal in less than two weeks and seldom develop infection complications. Counter-intuitively, sloths can also swim quite well. The sloth's diet is mainly leaves and buds, and they forage and live alone. The sloth's home range can be up to 6.5 hectares (16 acres), and part of their home territory is usually inherited from the mother.

Species name:	*Bradypus torquatus*
Features:	Proportionately small face; large clawed limbs; rough coat coloured with brown, tan, grey
Habitat:	Rainforests
Distribution:	Eastern Brazil
Length:	Up to 50cm (20in) without tail
Weight:	Up to 4kg (8lb 12oz)
Breeding:	One offspring after a six-month gestation

XENARTHA: DASYPODIDAE

Nine-banded Armadillo

The nine-banded armadillo is so-called from the number of flexible bands around the middle of its hard, scaly outer 'armour', which is formed from ossified dermal plates. Only a few species of armadillo (particularly the three-banded armadillos) roll themselves up into balls when attacked by a predators. The nine-banded is not one of them; instead it tries to outrun the predator – it can move at surprising speed – or flees into a burrow and uses the outer carapace to wedge itself securely inside. Armadillos have piglike snouts and an exceptional sense of smell. They feed on a wide range of plant and animal foods, which are detected by snuffling the nostrils close to the ground. The bulk of the diet is made of insects, particularly ants and beetles, with lizards, frogs and other small creatures also taken. Nine-banded armadillos establish home ranges, centred on burrows, but these are not defended on a territorial basis.

Species name:	*Dasypus novemcinctus*
Features:	Bony outer carapace with 8–10 flexible bands; sand-coloured body; large ears; blunt snout
Habitat:	Grasslands and forests
Distribution:	Southern USA down through South America
Length:	Up to 57cm (22.4in) without tail
Weight:	Up to 6.5kg (14lb)
Breeding:	Four offspring after a 120-day gestation

XENARTHA: MYRMECOPHAGIDAE

Silky Anteater

The silky anteater has a slothlike appearance and a similar slow movement, and unlike other anteaters it is a primarily arboreal creature, inhabiting forests throughout Central and northern South America. Each foot has a large, hooked claw for gripping onto branches, and the tail is fully prehensile, with a bare gripping pad near the tip. Silky anteaters are mainly found in ceiba trees, as the silvery fibres around the seed pods give the anteater's silver-grey skin an ideal camouflage background, shielding the anteater from predators such as birds of prey. In addition, the anteater gains further defence by being mainly nocturnal. The Silky anteater subsists almost entirely on ants and termites; it tears open the nests with its claws and licks up the insects with a long tongue (it can easily eat 8000 ants per day). Both males and females work in the raising of the young, with the male sharing in carrying the infant and in providing regurgitated food.

Species name:	*Cyclopes didactylus*
Features:	Silver-grey fur; brown stripe on chest; clawed limbs; prehensile tail
Habitat:	Forests
Distribution:	Central and northern South America
Length:	Up to 21cm (8.3in) without tail
Weight:	Up to 275g (10oz)
Breeding:	One offspring after a 120- 150-day gestation

XENARTHA: MYRMECOPHAGIDAE

Giant Anteater

The giant anteater has a long, thin snout that grows up to 45cm (18in), from which protrudes an elongated, sticky tongue that in itself can reach up to 60cm (24in) from the mouth. These physical attributes enable the giant anteater to eat several thousand ants or termites in a few minutes; the tongue also features tiny rearward-pointing spines to drag out the insects and larvae. The giant anteater tears insect nests open with its large front claws, and it keeps these in good condition (as well as protecting its own feet pads from the claws) by walking mainly on its knuckles. Giant anteaters are solitary, wandering creatures. Their home ranges can be as large as 25 sq. km (10 sq. miles), but these are not territorially defended, and encounters with other adults are generally non-violent. The giant anteater's main defence involves standing on its rear legs and slashing out with its front claws. Giant anteaters can dig well, but do not construct tunnels, sleeping instead in burrows abandoned by other animals or in rocky crevices or areas of dense vegetation.

Species name:	*Myrmeophaga tridactyla*
Features:	Grey and black fur with white markings; long bushy tail; extremely long, thin snout; tiny ears
Habitat:	Forests and grasslands
Distribution:	South America
Length:	Up to 2m (6ft 6in) without tail
Weight:	Up to 39kg (86lb)
Breeding:	One offspring (rarely two) after a 190-day gestation

XENARTHA: MYRMECOPHAGIDAE

Southern Tamandua

The southern tamandua is another anteater species of South America. It has all the classic anteater features: long snout, slender head, powerful clawed limbs. Typical habitats are forests and savannah, and in arboreal environments the southern tamandua is a capable climber – up to 60 per cent of its time is spent in the trees, where it will collect various insects (usually tree-dwelling ants and termites) with its long, sticky tongue. It will also raid bees' nests for their honey. Southern tamanduas are nocturnal creatures, and with poor eyesight they have to rely upon their acute senses of smell and hearing. They have few predators apart from indigenous big cats and domestic dogs. If attacked, the tamandua puts its back against a tree or rock, grips onto foliage with its hind legs and stands up, ready to slash the attacker with its fearsome claws. Southern tamanduas are endangered by hunting, their leg tendons often being used to make rope.

Species name:	*Tamandua tetradactyla*
Features:	Yellow fur with black sections over shoulders, back and flanks; long slender tail; long, thin snout
Habitat:	Forests and grasslands
Distribution:	South America
Length:	Up to 88cm (35in) without tail
Weight:	Up to 8.5kg (19lb)
Breeding:	One offspring after a 130- to 150-day gestation

Glossary

Agile: The ability to move quickly and gracefully.

Blubber: Fatty deposits found under an animal's skin that act as insulation.

Breeding: To produce offspring by hatching or gestation.

Boss: A rounded projection in the middle of the head of some ungulates where the horns or antlers meet.

Browser: An animal that feeds mainly on vegetation above the ground, such as leaves and branches.

Buck: The male of certain species of horned animals, mainly deer and antelope.

Bull: A large male animal; a term particularly applied to bovine animals and whales.

Carnivore: An animal that has a diet of meat.

Colony: A community of animals.

Cow: A female bovine animal or the females of several other species, including elephants and dolphins.

Crepuscular: Relating to the hours of dawn and dusk.

Diet: The type and amount of food and drink that is regularly consumed.

Delayed implantation: A reproductive state in which the fertilized embryo floats free in the uterus without developing, sometimes for a period of several months, before embedding in the uterine wall and beginning its 'true' gestation.

Doe: Female deer.

Dorsal fin: A fin located on the backs of fish and aquatic mammals, used mainly to control direction.

Echolocation: The use of sound-wave emissions to detect objects and shapes, used by certain species of animals including bats and dolphins.

Evolution: A change in the physical characteristics of animals which is believed to happen over time.

Family: In the system of animal classification (taxonomy) animals are split into groups based on biological similarities. For every animal there are seven distinct groupings: Kingdom, Phylum, Class, Order, Family, Genus and Species.

Frugivore: An animal that lives off fruit.

Genus: A biological grouping between species and family, denoting animals with common characteristics.

Gestation: The development of an animal within its mother from conception to birth.

Grazer: An animal that feeds on grasses and ground-dwelling plants.

Habitat: The natural home environment of an animal or plant.

Herbivore: A plant-eating animal.

Hibernation: An animal's self-imposed period of biological dormancy, usually undertaken during the winter months.

Host: Refers to an animal that is home to parasites.

IUCN: International Union for the Conservation of Nature and Natural Resources, otherwise known as the World Nature Union.

Keratin: A fibrous protein material occurring in the formation of structures such as hair, nails, horns, hooves, feathers and claws.

Lodge: The home of a beaver constructed out of branches and mud.

Moult: To shed hair or skin to allow for new growth.

Nocturnal: Active in the night-time hours.

Native: Originating from a particular place or vacinity.

Omnivore: An animal that eats both meat and plant material

Order: A biological classification of animals between class and family.

Parasite: An animal that obtains food or other such benefits by living off the body of another animal.

Salt lick: A patch of ground that animals lick to obtain salt.

Territorial: In the animal kingdom, groups or individuals will often fight to defend their territory from intruders.

Ungulate: A hoofed mammal.

Vertebrate: An animal with a backbone.

Vixen: A female fox.

Warm blooded: Animals that produce their body heat through metabolic processes and maintain a core temperature within a constant, narrow range. All mammals are warm-blooded.

Index

Page numbers in **bold** refer to main entries

aardvark 69, **306**
addax **15**
aestivation 173
African buffalo **44**
African bushpig **70**
African civet **143**
African elephant 10, **267**
African lion 100
African porcupine 282
African wild ass **208**
African wild dog **80**, 95
aguoti **275**
alpaca **51**
Alpine ibex **23**
Alpine marmot **298**
American badger **129**
American bison **19**
American black bear 138
American buffalo 13, 19
American mink 124, **126**
anteater 196, 207, 217, 309–11
antelopes 14, 16, 29–38, 45–7, 88, 105
antlers 53–9, 61–2
  see also horns
apes 261–4
Arctic fox **73**
Arctic ground squirrel **302**
Arctic hare **183**

armadillo 217, **308**
Artiodactyla 14–72
Asian elephant **266**
Asian Red Dog **79**
Asiatic black bear **141**
Asiatic lion **100**
Asiatic Wild Dog **79**
ass 208
Australian false vampire bat **160**
axis deer **54**, 79
aye-aye **252**

babakoto **255**
babirusa **67**
baboons 248, 251
Bactrian camel 48
badgers 81, 119, 120, 129
bald uakari **236**
baleen whales 8, 149, 150, 157
bamboo 134
banded anteater **196**
banded mongoose **148**
bandicoots 197, 204
bank vole **284**
bantengs 79
bat-eared fox **82**
bats 12, 160–8
bears 13, 73, 103, 134–41, **158**
beavers 12, 271–2, 290
bees 120
beluga **158**

Bengal tiger 103, **104**
big brown bat **166**
bighorn sheep 40
bilby **204**
binturong **142**
bipedalism 141, 262
bison 19
black bear **138, 141**
black-capped monkey **237**
black panther 102
black rat **293**
black rhinoceros **213**
black-tailed jackrabbit **184**
black-tailed prairie dog **296**
blackbuck **18**
blesbok **26**
blubber 140
blue-backed jackal 74
blue whale 7, 13, **150**
bobcat **94**
bonobo **262**
bontebok **26**
bottlenose dolphin **156**
Bovidae 15–47
brachiation 253, 254
Brazilian tapir **216**
brown bear **139**
brown capuchin **237**
brown hyena **110**
brown-mantled tamarin 233

brown rat **292**
brush-tailed possum **202**
buffalo 13, 19, 44, 102, 105
Burchell's zebra **209**
burrows 11, 69–70, 74, 85, 148, 205
bush duiker **43**
bushbaby **260**
bushpig **70**

California sea lion **221**
camels 48–52
camouflage 71, 73, 92, 158, 217, 309
Cape buffalo **44**
Cape porcupine **281**
capybara 101, 280
caracal **88**
caribou **61**, 77
Carnivora 11, 73–148
cat family 87–106
cattle 15–47
cavy 273, 275
Cetacea 149–59
chamois **41**
cheetah **87**, 95
chimpanzee **263**
chinchilla **274**
chipmunk 118, **303**
Chiroptera 160–8
chital **54**, 100, 104
chousingha **46**
civet **143**, 171
clouded leopard **99**
coat
  for cold weather 21, 35, 39, 61, 73
  human exploitation of 50–2, 104, 192, 204
  sea creatures 112, 219
collared peccary **72**
colobus monkey **241**
colugo **169**
common cuscus **201**
common dolphin **152**
common duiker **43**
common genet **144**
common marmoset **230**
common seal **229**
common tenrec **181**
common wombat **205**
conservation 52, 115
coterie 296
cotton-top tamarin **234**
cottontails 129
cougar **89**
coyote **76**, 84
coypu **270**
crabeater seal 224
critically endangered species 15, 135, 157, 212, 227, 232, 244, 256
see also endangered species
crocodiles 25, 104
crop destruction 278, 295–6
  bears 137, 141
  rabbits 185
Cuban hutia **269**
cuscus **201**

deer 14, 53–62, 79, 89, 93–4, 99–101, 103
deforestation 13, 118, 134, 232, 236–7, 240, 252, 264
'delayed implantation' 115, 118, 119, 130, 168
Demarest's hutia **269**
dens 12, 69, 83–5, 95, 122
desert hedgehog **173**
desert jerboa **276**
desert lynx **88**
desman **179**
dhole **79**
Diana monkey **240**
dik-dik **34**
dingo 191
dog family 73–86
dolphins 152, 155–6
domesticated mammals 21, 45, 48–51
see also farm livestock
dormouse **279**, 294
douc langur **249**
'drays' 200
dromedary 48, **49**
duckbill platypus 9, **206**
dugong 304
duiker **43**
dwarf mongoose **146**

ears 7, 70, 80, 82, 84, 86, 92
eastern chipmunk **303**
eastern grey kangaroo **190**

315

eastern grey squirrel **300**
eastern spotted skunk **128**
echidna 9, **207**
echolocation 151, 158–62, 165–6, 175
edible dormouse **279**
eland **45**
elephant seal **130, 226**
elephants 10, 12, 104, 266–7
elk **53**
emperor tamarin **233**
endangered species 11,13
   Arteriodactyla 28, 48, 52
   Carnivora 78–9, 91, 99, 124
   Equidae 208, 210
   Insectivora 174
   marsupials 189, 196
   primates 234, 236, 240, 252, 257, 259, 261, 264
   rhinoceros 211
   rodents 282, 301
   Sirenia 304
   whales 149
   see also critically endangered species; extinct species; vulnerable species
ermine 122
Eurasian badger **119**, 129
Eurasian otter 115
Eurasian pine marten **117**
Eurasian red squirrel **301**
Eurasian shrew **177**
Eurasian water shrew **176**
European beaver **272**
European fox 196
European hedgehog **172**
European mink **124**
European mole 175, **180**
European polecat **123**
European rabbit **185**
European wildcat **96**
extinct species 13, 52, 227

fallow deer **56**
farm livestock 89, 100, 101, 123, 163
   see also domesticated mammals
fat dormouse **279**
fennec fox 84, **86**
ferret 123
field vole **288**
fish-eating bat **162**
fisher **118**
flying lemur **169**
forest duiker 43
four-horned antelope **46**
foxes 73, 78, 81–6, 196
frogs 95
fur see coat
fur seals 219–20, 224

gaur **20**
gazelles 18, 27, 87

gelada baboon **251**
gemsbok **37**
genet 144–5
genetic diversity 134
Geoffroy's cat **90**
gerbils 92, 285–6
gerenuk **33**
ghost bat **160**
giant anteater **310**
giant forest hog **68**
giant otter **127**
giant panda 13, **134**
gibbons 253–4
giraffe **63**, 64, 102
global warming 228
glutton **113**
goats 23, 35, 41
golden jackal **75**
golden lion tamarin **232**
gopher **278**
gorillas **261**
grasshopper mouse **291**
greater bulldog bat **162**
greater horseshoe bat **165**
grey fox **83**
grey seal **223**
grey squirrel **300**
grey whale **157**
grey wolf **77**
Grimm's duiker **43**
grizzly bear **139**
grooming 8–9
groundhog **299**
group structure 12

guanaco **50**, 51
'gulpers' 150

habitat destruction 15, 20, 160, 203–4, 212, 274
  carnivores 78, 104, 131
  moles 179
  primates 239, 256–7
  see also crop destruction; deforestation
hair 6–7, 8–9
  see also coat
hairy rhinoceros **212**
hamadryas baboon **248**
Hanuman langur **250**
harbour seal **229**
hares 182–4
harp seal **228**
harvest mouse **287**
hazel dormouse **294**
hedgehogs 9, 172–3
herbivores 11
hibernation 9, 81, 138, 167–8, 172, 181, 279, 294, 299, 302–3
hippopotamus 65, **66**
holts 127
honey badger **120**
hooded seal **222**
horns 14–18, 20–33, 35–42, 44–6
  see also antlers
horseshoe bat **165**
house mouse **289**

human predators 13, 100, 189, 192, 197, 243, 252
  see also habitat destruction; hunting
humpback whale **151**
humps 48–9
hunting 13, 115, 197
  Arteriodactyla 18–21, 24, 26, 40, 52, 56, 71
  carnivores 90, 93, 95, 97
  primates 236, 240
  rhinoceros 212–13
  rodents 280, 282, 301
  seals 223, 227–8
  whales 158–9
Huon tree kangaroo **189**
hutia **269**
hyenas 7–8, 69, 78, 95, 108–10
hyrax 170–1

Iberian lynx 91
ibex 23, 28
impala **16,** 87, 88
Indian flying fox **164**
Indian muntjac deer **58**
Indian rhinoceros **214**
indri **255**
Insectivora 172–81
intelligence 133, 153, 156, 263

jackals 74–5
jackrabbit **184**
jaguar **101**, 216
jaguarundi **98**

Japanese macaque **242**
javelina **72**
jaws 7–8, 51, 101, 104, 113, 187
jerboa 92, **276**
jird 286

kangaroo 10, 188–90, 193
killer whale **154**, 158
kinkajou **132**
kit fox **84**
klipspringer **36**
koala 200, **203**
kob 32
krill 224

langur 249–50
lar gibbon **253**
large-spotted genet **145**
lek 32
lemmings 96
lemurs 169, 255–9
leopard seal **224**
leopards 95, 99, 101, **102**, 106, 171
lesser bushbaby **260**
lesser horseshoe bat 165
lesser panda **131**
lion-tailed macaque **244**
lions 100, **105**, 108
  as predators 25, 69, 110, 120
lizards 95, 96
llama 50, 51, 52
lobtailing 151
lodge 12, 271–2

long-nosed bandicoot **197**
lowland gorilla **261**
lynx 88, 91, 103

macaques 242, 244–5
mainland serow **24**
Malayan tapir **215**
mammal characteristics 6–8
manatee **305**
mandrill **246**
maned three-toed sloth **307**
maned wolf **78**
mara **273**
margay **97**
marmosets 230–1
marmots **298**
marsupials 9, 11, 186–205
meerkat **107**
Mexican free-tailed bat 12, **161**
mice 6, 10, 287, 289, 291
migration
 Arteriodactyla 25, 27, 42, 53, 61
 seals 220, 226, 228
 whales 157–8
mink 124, 126
mobs 190, 193
mole-rat 6, **268**
moles 178–80, 278
Mongolian gerbil 286
mongoose 107, **146–8**
mongoose lemur **257**

monk seal 11, **227**
monkeys 93, 101, 104, 230–51
Monotremata 206–7
moose **53**, 77
mountain goat 35
mountain gorilla 261
mountain lion **89**
mountain zebra **210**
mouse-eared bat **167**
mule deer 59
muntjac deer **58**
musk 175
musk deer **57**
musk ox **39**
muskrat **290**
Mustelidae 111–29

naked mole-rat 6, **268**
nests 11
night vision 84, 86, 103, 133, 252, 265
nilgai **22**
nine-banded armadillo **308**
North American beaver **271**
North American porcupine **277**
North American river otter **115**
Northern fur seal **220**
northern lynx **91**
northern pygmy gerbil **285**
northern right whale **149**

northern short-tailed shrew **175**
nose 42
 see also snout
noseleaf 165
nuchal crest 211
numbat **196**
nyala **47**

ocelot 93, 97
okapi **64**
omnivores 11
orang-utan **264**
orca **154**, 158
oribi **38**
ossicones 63, 64
otters 6–7, 8–9, 112, 115, 127
oxen 39, 104

pademelon **195**
pandas 13, 131, 132, 134
pangolin **217**
panther **89**
Parry's wallaby **191**
patagium 169, 297
Patagonian cavy **273**
pedicels 58
perfume 57
Perissodactyla 208–16
pest control 95, 145, 161, 165, 166
pests 164, 185, 278, 293, 300
pets 98, 174, 203, 234, 259, 274, 286

see also domesticated mammals
pigs 67–71
pilot whale **153**
pine marten 117, 118
Pinnipedia 218–29
pipistrelle bat **168**
plains zebra **209**
pocket gopher **278**
pods 154, 158
polar bears 13, 73, **140**, 158
polecats 114, 123
pollution 13, 115, 124, 179, 304
porcupines 7, 99, 118, 277, 281–2
possums 198, 200, 202
poultry 78, 98, 123, 145
prairie dog **296**
primates 12, 13, 230–65
proboscis monkey **247**
promisians 252
pronghorn **14**
'pronking' 17, 27
pudu **60**
puma **89**, 216
pygmy chimpanzee **262**
pygmy hippopotamus **65**
pygmy marmoset **231**
pygmy rock mouse 6

quoll **186**
quolla **194**

rabbit-eared bandicoot **204**
rabbits **185**
raccoon 131–2, **133**
raccoon dog **81**
ratel **120**
rats 6, 268, 292–3
red deer 59
red fox 85
red howler monkey **235**
red kangaroo **193**
red-necked wallaby **192**
red panda **131**, 132
red squirrel **301**
red uakari **236**
reindeer **61**
reproduction 9–10
rhesus monkey **243**
rhinoceros 104, 211–14
right whale **149**
ring-tailed lemur **258**
ring-tailed possum **200**
river otter 115
roan antelope **29**
rock hyrax **171**
rock mouse 6
rock rabbit **170**
Rocky Mountain Goat **35**
rodents 10, 11, 268–303
  see also pest control
roe deer **55**
'rookeries' 223
ruffed lemur **259**
Russian desman **179**
rusty-spotted genet **145**

sable antelope **30**
sacred animals 164, 250
saiga **42**
sambar **62**
Sambar deer 100
sand cat **92**
scaly anteater **217**
sea lion **221**
sea otter 6–7, 8–9, **112**
seals 11, 12, 130, 140, 219–29
serow **24**
serval **95**
sett 119
sheep 40
ship rat **293**
short-nosed echidna **207**
short-tailed vole **288**
shrews 174–7
siamang gibbon **254**
Siberian musk deer **57**
Siberian tiger **103**
side-striped jackal **74**
sifaka **256**
silky anteater **309**
silverback 261
skin 7
skunk 114, 121, 128, 198
skyhopping 151
sloth **307**
sloth bear **136**
small-spotted genet **144**
snakes 92, 120, 147, 171
snout 68, 178, 179, 215
snow leopard **106**
snowshoe hare **182**

319

solenodon **174**
South American tapir **216**
southern flying squirrel **297**
southern fur seal **219**
southern tamandua **311**
spectacled bear **137**
speed 14, 69, 76, 87
sperm whale 6, **159**
spermaceti oil 159
spines 172–3, 277, 281–2
spiny anteater **207**
spotted dolphins **155**
spotted hyena **108**, 110
spotted skunk **128**
springbok **17**, 27
springhare **295**
squirrel monkey **239**
squirrels 297–302
star-nose mole **178**
stoat **122**, 125
striped hyena **109**
striped polecat **114**
striped possum **198**
striped skunk **121**, 128
sugar glider **199**
Sumatran rhinoceros **212**
sun bear **135**

tahr **28**
tail, prehensile 132, 142, 200–1, 235
tamandua **311**
tamarins 232–4
tapir 42, 215–16
tarsier **265**

Tasmanian devil **187**
tayra **111**
teeth 8, 57, 72
 see also tusks
Temminck's pangolin **217**
tenrec **181**
termites 136, 146, 196
Thomson's gazelle **27**
tigers 13, 62, 79, 103, 104
tongue 132, 135
tree fox **83**
tree hyrax **170**
tree kangaroo **188**
trunks 130
tusks 58, 67–70, 218

uakari **236**
vampire bat **163**
Verreaux's sifaka **256**
vicuña 51, **52**
vocalisation 150–1, 153, 156, 158–9
 antelopes 47
 carnivores 76–7
 primates 171, 235, 240–1, 244, 254, 263
voles 283–4, 288
vulnerable species 141, 160, 204, 274, 295
 Arteriodactyla 18, 20, 26, 50, 65
 carnivores 77, 90, 97
 see also endangered species

wallaby 188, 191–2, 194–5

walpurtis **196**
walrus **218**
'wanderoo' **244**
warthog **69**
water buffalo 105
water shrew **176**
water vole **283**
waterbuck **31**
weasels 98, 111–29
Weddell seal **225**
West Indian manatee **305**
western grey kangaroo 190
whales 6, 7, 8, 13, 149–59
whip-tailed wallaby **191**
white-handed gibbon **253**
white rhinoceros **211**
white-tailed deer **59**
white-tufted-ear marmoset **230**
wild boar **71**, 79, 103, 104
wildcats **96**
wildebeest 10, **25**, 80, 87, 105, 108
wolverine **113**
wolves 77–8
wombat **205**
woodchuck **299**
woolly monkey **238**

yak **21**, 106

zebras 80, 105, 108, 209–10
zoos 52, 103, 156
zorilla **114**